Poisonous Plants
and Animals

In the same series

Chatto Nature Guides

Poisonous Plants and Animals

Illustrated and identified with colour photographs

Horst Altmann
Translated and edited by
Gwynne Vevers

Chatto & Windus · London

Published by
Chatto & Windus Ltd.
40 William IV Street
London WC2N 4DF

          *

Clarke, Irwin & Co Ltd.
Toronto

British Library Cataloguing in Publication Data

Altmann, Horst
    Poisonous plants and animals. —
        (Chatto nature guides).
    1. Poisonous plants — Europe — Identification
    2. Poisonous animals — Europe — Identification
    I. Title
    581.6'9'094   QK100.E9
ISBN 0 7011 2525 X (hb)
ISBN 0 7011 2526 8 (pb)

Title of the original German edition:
Giftpflanzen · Gifttiere
© BLV Verlagsgesellschaft mbH, München, 1979
English Translation © Chatto & Windus Ltd 1980

Printed in Germany

Contents

Introduction

Every now and again the newspapers draw attention to a case of poisoning by plants or animals. In the majority of cases the victim has been a child who was attracted by the bright berries of a plant in the garden or out in the country. This book is intended as a guide to help in the identification of the poisonous plants and animals that are likely to be encountered in Britain and Europe.

Fortunately, cases of poisoning are rare but, as will be seem from the descriptive text, they may be caused by a wide variety of different substances. In those cases in which the symptoms described are serious the victim should be taken immediately to the nearest doctor or hospital. In cases of poisoning by fungi or flowering plants it is quite essential that samples of the plant known to be or thought to be involved should also be taken along so that the doctor can know the type of poison he is dealing with.

Poisonous plants

The poisonous plants illustrated and described here belong to two very different groups, namely the fungi and the flowering plants. Fungi are plants without the green pigment chlorophyll, which are either parasites, deriving their nutrients from living organisms, or saprophytes that live on decaying organic matter. The majority are saprophytes.

The terms mushroom and toadstool are attractive, but useless for purposes of identification. Some people use the term mushroom for edible fungi and toadstool for those that are inedible either because they are tough and unpalatable or because they are poisonous. This is a dangerous separation for in a group of closely related fungi some may be edible, others poisonous. This happens, for instance, in the genus

Amanita (pp. 16-22), in which most species are poisonous, some deadly poisonous, while at least one species is edible. The only safe way of dealing with fungi is to know the exact scientific name. Popular names are quaint and sometimes frightening, e.g. Death Cap (p. 16), Destroying Angel (p. 19) and the Sickener (p. 26), but they sometimes vary in different parts of the country. Furthermore, popular names are useless internationally. It would be unproductive to tell a doctor on the Continent that you think you have been poisoned by the Death Cap, but if you say *Amanita phalloides* he will know at once what you mean. In practice a very small percentage of fungi are poisonous, but those that are can produce the most unpleasant and sometimes fatal symptoms.

The flowering plants contain chlorophyll, the green pigment which is intimately concerned with photosynthesis. This is the process whereby, in the presence of light, water, nutrient salts and chlorophyll, a plant can synthesize its own tissues and grow. Only green plants can do this and they provide the basic food for animals.

The flowering plants range in size from the duckweeds a few millimetres across to the largest forest trees. Here again, only a few are dangerous to man, and many of these have been exploited by man to produce valuable medicinal products when used in the correct, usually minute quantities (e.g. atropine, digitalin).

Poisonous animals

In some parts of the world there are animals which are poisonous to eat. For instance, some of the fishes found in tropical seas are poisonous at certain times of the year or at all times. Certain of the pufferfishes have poisonous flesh, yet they are eaten, after special preparation, as a delicacy in Japan.

In Europe, however, the poisonous animals are not of this type. Instead they are mostly animals which administer a poisonous bite or sting. Such animals belong to a wide variety of animal groups.

Among the invertebrates or animals without backbones there are poisonous representatives among the scorpions, spiders, ticks and insects.

Scorpions use the sting at the end of their tail to deliver a dose of venom to their prey, but they only do this when the prey puts up a fight. Spiders bite with a pair of front limbs known at the chelicerae. These inject venom into the prey which is usually an insect. When sucking blood ticks may transmit virus diseases.

Insects have different ways of administering their poison. Mosquitoes, gnats and horseflies have their mouthparts modified to form a piercing organ capable of penetrating human and other skin. When this happens the insect's saliva passes into the wound and it is the substances contained in this saliva which cause pain, swelling and sometimes more serious symptoms. The true bugs also have a puncturing organ used for sucking either plant saps or blood. Those that suck blood also cause pain when their saliva enters the victim's skin.

Among the insects, however, the most sophisticated method of poisoning is that used by the bees and wasps. Here the females, but never the males, have a sting at the rear end which is actually a modified ovipositor or egg-laying organ. This is supplied with a venom gland. When the very sharp sting enters the skin the gland delivers a dose of the venom which is responsible for the subsequent symptons.

The vertebrates or animals with backbones contain relatively few poisonous animals. The birds, for example, have no poisonous representatives, and the mammals only one or two, e.g. the Duck-billed Platypus has venomous spurs. The toads, frogs and salamanders produce poisonous substances in their skin but these only affect humans who handle them and then transmit these secretions to delicate parts of their own body, e.g. the mucous membrane of the eyes, nose and mouth.

If the bees and wasps are the Borgias of the invertebrate world then the snakes are their counterparts among the vertebrates. Venomous snakes are found mostly in the tropics and subtropics. These include cobras with representatives in southern Asia and Africa, the rattlesnakes of America and the boomslang, mambas and puff adders of Africa. Australia, too, has its own group of venomous snakes, some deadly. The waters of the Indian and Pacific Ocean also have a number of very dangerous sea-snakes which are responsible for many deaths every year among fishermen in south-east Asia.

Compared with the relative abundance of venomous snakes in warmer regions, Europe has very few and Britain only one, the Adder or Common Viper. In fact, all the European venomous snakes belong to the viper family, known scientifically as the Viperidae.

8

Plant and Animal Poisons

Poisons are substances which have a toxic action on the normal cells of an animal. Some may merely irritate the skin or more seriously the mucous membrane of the eyes, nose and mouth. Several instances of such substances will be found in the descriptive texts.

Other poisons have a toxic effect when eaten and these are primarily the plant poisons. The chemical composition of plants is a highly complex subject. Food plants such as cereals, pulses, peanuts and sunflower seeds are well known for their content of proteins, carbohydrates and fats. Others have resinous compounds, e.g. the pines and spruces. In most poisonous plants the toxic agents are glycosides or alkaloids. These substances occur very commonly among plants and in many cases they are quite harmless. Some glycosides and alkaloids are used in medicine, the dose being very small and controlled. Thus, the familiar Foxglove yields a number of glycosides of which digitoxin and digoxin are used in very small amounts, usually by mouth, in the treatment of heart failure. Similarly, the less familiar Deadly Nightshade, containing a number of alkaloids that withstand drying and boiling, yields the drugs atropine and scopolamine which are used, for instance, to inhibit salivary secretion before anaesthesia.

In certain instances poisonous plants have an obnoxious smell or an unpleasant taste and these are less likely to be eaten. In many such plants the poisonous substances may be present in all parts, in others only in the leaves, seeds or roots. The concentration of poison in the plant tissues also varies considerably. Many plants are only poisonous when eaten in large quantities and in the raw state. On the other hand, very small quantities of certain fungi can cause serious trouble, and this applies particularly to the appropriately named Death Cap. A student out in the country studying the fungi once collected a single specimen of this fungus and examined its structure before throwing it away. One hour later he ate his sandwiches without washing his hands. That night he suffered severe vomiting and diarrhoea, and did not fully recover until about two days later.

The chemical composition of many animal poisons such as those of spiders is not known in any great detail, but they are usually protein in nature. More research has been done on the venoms of snakes, because they are such a serious hazard to humans. The venom injected into the skin by a snake is in fact just saliva, but a particularly potent saliva, produced in the salivary glands situated in the upper jaw. These venoms contain several different poisonous substances, and they have different actions on the victim's body. The venom of cobras and some other snakes acts on the nervous system and in severe cases this may cause respiratory failure. Other venoms affect the tissues in the area of the bite and they may cause damage to the blood.

In general, the victims of poisoning, whether by plants or animals, should be taken immediately to a medical doctor or a hospital. In lay hands the induction of vomiting is not now recommended. Nor is the old first aid treatment for snake bite, which involved cutting the skin and sucking out the venom or rubbing the bitten area with permanganate. These procedures waste time and may cause unnecessary damage.

Whenever possible the victim should be taken to the doctor or hospital together with the fungus, flowering plant or snake thought to be responsible for the poisoning.

Glossary of technical terms

alkaloids	nitrogenous compounds, often poisonous
anaphylactic	an agent that decreases immunity instead of increasing it
antispasmodic	an agent which relieves spasm in muscles
ataxia	loss of control of lib movements
bradycardia	slowing of the heart beat, as detected by the pulse rate
cardiac	of the heart
choleretic	an agent which stimulates output of bile
cultivars	cultivated varieties of plants
cyme	of an inflorescence which has several flowering branches, the centre flowers opening first
dermatitis	inflammation of the skin
diuretic	an agent which stimulates output of urine
entire	of leaves in which the edge is not indented or toothed
glycosides	compounds containing a carbohydrate molecule. Frequent in plants and often poisonous.
inflorescence	a flowering branch with one or more flowers
opposite	of two leaves growing from the same level on opposite sides of the stem
panicle	of an inflorescence with the axis divided into branches each bearing two or more flowers.
photophobia	abnormal intolerance of light
raceme	of an inflorescence in which the flowers are borne on a single individed axis
sessile	of leaves or flowers without stalks
spadix	a fleshy axis bearing numerous sessile flowers
spathe	a leaf-like bract enclosing the inflorescence in certain plants

thready pulse	one that is scarcely perceptible
umbel	of an inflorescence in which several branches start from the same point (like an inverted umbrella), each branching bearing one or more flowers.
vasomotor	affecting the calibre of the small blood vessels

Poison Centres in the British Isles

Advice on poisoning by plants and animals
can be sought at the following:

ENGLAND

National Poisons Information Centre
New Cross Hospital,
Avonley Road, London, S.E.14.
Telephone: 01-407-7600.

Walton Hospital,
Rice Lane, Liverpool L9 1EA
Telephone: 051-525 3611
in conjunction with the
Liverpool School of Tropical Medicine,
Pembroke Place, Liverpool L3 5QA
Telephone: 051-709 7611.

SCOTLAND

Regional Poisoning Treatment Centre,
Royal Infirmary,
Lauriston Place,
Edinburgh 3.
Telephone: 031-229-2477.

WALES

Poisons Information Centre,
Royal Infirmary,
Cardiff.
Telephone: 0222-492233.

**REPUBLIC
OF IRELAND**

Poisons Information,
Jervis Street Hospital,
Dublin.
Telephone: Dublin 745588.

Poisonous Plants
and Animals

Lorchel   *Gyromitra esculenta*

Helvellaceae

Characteristics: the cap of this fungus is about the size of a fist, usually rather broader than tall and its thick convolutions are reminiscent of a brain. Externally it is red-brown to blackish-brown, but the flesh is white and waxy with a pleasant smell and taste.—**Distribution.** Europe, from April to May on calcareous and sandy soils in coniferous and mixed woods, also among felled trees, extending up into the hills.—**Poison:** gyrometrin and helvellic acid. Serious poisoning occurs after eating this fungus uncooked, sometimes leading to death. Even after cooking poisoning sometimes occurs, because it is possible that not all the poisonous substances have been destroyed. Symptoms may occur 2-4 hours after a meal or even as much as 12 hours later. They include vomiting and diarrhoea, jaundice, convulsions, loss of consciousness and abdominal pain. After 2 days to 2 weeks the patient may die due to liver failure. The eyes and skin may be irritated if the fungus is handled. It is eaten in some places after being well cooked, but the water in which it is boiled must be discarded.

Handsome Clavaria   *Ramaria formosa*

Clavariaceae

Also known as *Clavaria formosa.*—**Characteristics:** a much branded pinkish to orange fungus with pink or lemon-yellow tips to the branches. The flesh is white and very fragile.—**Distribution:** Europe, mainly in deciduous woodland, particularly beech, between August and October, but not common.—**Poison:** not yet identified. Symptoms include abdominal pain, vomiting and diarrhoea. This fungus is sometimes said to be edible if the yellow tips to the branches are removed, but it has an acrid taste, although the smell is quite pleasant.

The similar, but paler, *Clavaria pallida* may cause abdominal pain and diarrhoea in as little as 15 minutes after consumption.

There are about 40 different species of *Clavaria* in Britain.

Death Cap *Amanita phalloides*

Agaricaceae

Characteristics: the cap is up to 15 cm across, at first spherical, later becoming slightly curved to flat. The surface of the cap has a satin-like surface when dry, but becomes rather sticky in wet weather. The colour is olive to grey-green but may be yellow, brownish or even whitish, and peels off easily. The flesh is white, almost tasteless and without any smell when fresh. The gills are also white with white spores. The stem is up to 15 cm tall, and always swollen at the base. This characteristic structure, known as the volva, may not be apparent unless the fungus is dug up. There is a large, whitish or greenish ring around the stem below the gills.—**Distribution:** Europe, mostly in deciduous woodland, more rarely in coniferous woods, and appearing from July to October.—**Poison:** phalloidin, phalloin, alpha-, beta- and gamma-amanitine. Phalloin is not heat-stable and loses its virulence when cooked. Phalloidin and the amanitine toxins are resistant to heat, desiccation and digestive enzymes. Amanitine is 10-20 times more poisonous than phalloidin. As little as 30-50 g of the fresh fungus (about 1 medium-sized cap) will be lethal to an adult and much less can kill a child. Phalloidin takes effect after a short latent period with intense gastro-intestinal symptoms. The action of amanitine appears 3-5 days after consumption of the fungus, causing degenerative changes in the heart, liver and kidneys and generalized muscle wasting. The poison is excreted mainly in the urine, but a small amount reaches the gall bladder. The first symptoms normally appear after 8-16 hours. In the phalloidin phase the victim suffers nausea, vomiting, diarrhoea like rice water, colicky abdominal pain, and cramp in the calf of the leg. This phase lasts 1-2 days, and is usually followed by a day of remission. In the second or amanitine phase, between the 3rd and 5th day, there may be signs of liver damage which may lead to jaundice and coma. There may also be damage to the kidneys. From about the 4th day, protein may occur in the urine, the blood sugar decreases, there is capillary damage to small blood vessels leading to splinter-like haemorrhages in the skin, damage to heart muscle and convulsions. The victim must be immediately taken to a hospital. The interval between consumption of the fungus and the start of the illness is important. If the latent

period is more than 5 hours the suspicion of poisoning by *Amanita* is likely. The most important task of lay persons is to take any remains of the fungus meal for thorough clinical examination, and all those partaking of the meal must also attend hospital. There is no doubt that this is the most dangerous of all fungi and it is reckoned that it causes at least 90 per cent of the recorded deaths by fungal poisoning. Furthermore, a very large percentage of cases of poisoning by this fungus are fatal. Some authorities say that 50 per cent of these cases result in death but this figure may be on the low side. In non-fatal cases the victim recovers very slowly, after two to three weeks of intense pain, and will require a long period of convalescence.

Destroying Angel *Amanita virosa*

Agaricaceae

Characteristics: the cap is at first conical, becoming spherical to arched as it grows: it is 5-10 cm across. The flesh and gills are white. The slender stem, 8-15 cm tall, is also white with a large white ring which may disappear with age. At the base of the stem is a large volva which is cup-shaped.—**Distribution:** Europe, usually in coniferous woodland, but also to be found in mixed woodland, often under beeches, from July to October.—**Poisons** and symptoms are in general as for the Death Cap. This is a deadly fungus with white flesh and a mild taste. The first signs of poisoning may appear after a latent period of at least 5 hours, and the victim must be taken immediately to a hospital.

Fly Agaric   *Amanita muscaria*

Agaricaceae

Characteristics: this is perhaps the best known of the poisonous fungi. The cap may reach a diameter of about 20 cm. Its yellowish to brilliant red skin, covered with numerous whitish warts, peels easily. The warts, which are often washed off by the rain, are the remains of the universal veil which surrounded the young fungus. The soft white flesh is almost tasteless.—**Distribution:** throughout Europe, appearing from summer to late autumn in deciduous and coniferous woods, in thickets and even under birch trees in moorland areas.—**Poison:** mycoatropine and muscimol, and also muscarine which is mostly concentrated in the skin of the cap. The use of the Fly Agaric as a stimulant was recorded as early as 1730. This fungus produces serious symptoms, but these are not often fatal. Evidently there are more cases of poisoning in southern Europe where this species has been confused with the edible *Amanita caesarea*, which is not uncommon in Italy and southern France. It has been reckoned that it would require the consumption of more than 10 Fly Agaric to kill an adult. The sensitivity of different individuals varies considerably and so does the content of poison in the fungus. The first symptoms almost always appear after 15-30 minutes, with salivation, nausea, vomiting and dilatation of the pupils. These are followed by signs of intoxication, cerebral stimulation, which may border on delirium (2-3 hours after consumption). Loss of consciousness may ensue, followed sometimes by death due to failure of respiration and an upset of the heart rate and rhythm.

Panther Cap   *Amanita pantherina*

Agaricaceae

Characteristics: the cap which is only about 10 cm in diamater, is at first hemispherical, and later flattish. The skin is pale brown to blackish-brown, becoming paler with age and is easily peeled. It is covered with the numerous scales or warts which are the remains of the universal veil, often arranged in rings. The flesh is white and tastes like radish.—**Distribution:** Europe, appearing fairly commonly from August to October, in deciduous and coniferous woods.—**Poison:** mainly muscarine (more than in the Fly Agaric), producing severe symptoms in 1-2 hours, which are in general similar to those caused by the Fly Agaric but more serious, with occasional deaths.

Red-staining Inocybe   *Inocybe patouillardii*

Agaricaceae

Characteristics: the cap, which has a diameter of 4-8 cm, is at first conical to domed, later becoming flattened. The colour of the cap is at first whitish, then yellowish to ochre-brown. When old or if touched or broken this fungus becomes brick-red. At first the white flesh has a mild taste, but with age this becomes unpleasant.—**Distribution:** Europe, in deciduous woodland, along the edges of woods and in grassy places; also in parks and gardens, appearing from June to September.—**Poison:** all parts contain muscarine, but 20 times as much as in the Fly Agaric. Deaths have occurred from eating this fungus which being completely white when young may be confused with the edible mushroom. From 15 minutes to 2 hours after consumption the victim suffers from headache, giddiness, sweating, lachrymation, nausea, abdominal pain, disturbances of vision, and constriction of the pupils. The eyes lose the capacity to focus near objects on the retina. The victim is fully conscious and rational, and does not show the cerebral irritation and excited behaviour seen in poisoning by the Fly Agaric. The pulse rate may decrease and death will be due to circulatory failure.

Livid Agaric   *Rhodophyllus simuatus*

Agaricaceae

Characteristics: the cap is ivory-coloured to pale brownish, becoming grey-brown with age. It is at first conical but as it grows it flattens out and reaches a diameter of up to 20 cm. The gills are at first yellowish, later becoming pink. The flesh is firm and white with a strong smell. The stem is 6-12 cm tall, often becoming hollow with age.—**Distribution:** Europe, where it is found from May to September in deciduous woods on loamy soils.—**Poison:** not yet identified. It acts primarily on the digestive tract, producing vomiting and diarrhoea very soon after consumption of the fungus. In some cases there may be damage to the liver. The victim should be taken to hospital immediately for gastric lavage. Fatal cases have been recorded.

Pardine Tricholoma   *Tricholoma pardinum*

Agaricaceae

Characteristics: a medium-sized fungus, at first domed but later flattening out to a diameter of 6-10 cm. The cap is covered with broad brownish to ash-grey scales which are arranged roughly like the tiles on a roof. The broad gills are off-white with a greenish-yellow tinge. The flesh is white and the stem is up to 3-5 cm in diameter.—**Distribution:** Europe, appearing from summer to autumn, mainly in beech woods on chalky soils, but also in conifer forests. A rather uncommon fungus.—**Poison:** not yet identified. It causes severe gastro-intestinal disorder with vomiting and diarrhoea, which may be prolonged. The symptoms are severe but not fatal. The victim should be treated in hospital.

Sickener *Russula emetica*

Lactariaceae

Characteristics: the brittle, thin-shelled cap is 5-10 cm across, at first hemispherical becoming domed to flattened and slightly concave in the centre. The brilliant pale red to purple red skin may fade to yellow or whitish; it is sticky in damp weather and is easily removed. The flesh, stem and gills are white. This fungus has a pleasant vegetable-like smell, but a sharp, burning taste.—**Distribution:** Europe, appearing from June to July to November in deciduous and coniferous woods, extending up into the mountains.—**Poison:** muscarine, which slightly affects the liver and kidneys. The victim suffers violent abdominal pain accompanied by diarrhoea and vomiting, particularly after eating the fungus in the raw state.

Russula queletii

Lactariaceae

Characteristics: the cap is at first hemispherical but as it grows it expands to a diameter of about 8 cm and becomes funnel-shaped; the coloration is purplish-red to lilac turning much paler with age. The whitish gills change to greenish when pressed. The stem is up to 7 cm long and 2 cm thick. This fungus differs from other red and edible species of *Russula* in having an extremely sharp taste.—**Distribution:** Europe, appearing from July to October, mainly in coniferous woods.—**Poison:** evidently not yet identified. The effects are not very serious. About 2-4 hours after consumption of the fungus there is nausea, followed by vomiting and diarrhoea, usually accompanied by stomach pains.

Woolly Milk Cap *Lactarius torminosus*
Lactariaceae

Characteristics: a medium-sized fungus with a relatively thin cap which is always felt-like or shaggy. Later this pale pink to reddish cap becomes flatter and often funnel-shaped. The gills are pale pink, and the flesh whitish.—**Distribution:** Europe, particularly in warm areas, appearing in August to October in open woodland and heaths, often under birch trees.—**Poison:** the milky sap probably contains a cell poison that damages the liver. Consumption of the raw fungus has on several occasions led to gastro-enteritis and death. The white, milky sap has a sharp bitter taste which burns the mouth and tongue when the fungus is eaten raw. After a latent period of 4-5 hours the patient vomits and suffers colicky abdominal pain and may have diarrhoea. The poison is destroyed by heat.

Involute Paxillus *Paxillus involutus*
Paxillaceae

Characteristics: cap up to 15 cm in diameter, at first slightly domed and rolled in at the edge, later becoming flattened and funnel-shaped. The ochre-yellow to red-brown skin is difficult to peel. The pallid gills immediately turn dark brown when pressed.—**Distribution:** Europe, appearing commonly from July to November in deciduous and coniferous woods, in grassy places and moorland.—**Poison:** a toxin that is destroyed by heat. When eaten raw this fungus causes serious intestinal disorder. It must therefore be well boiled or roasted. It is said that people who have correctly prepared and eaten it over a period of years may develop an allergic-type reaction to it.

Devil's Boletus *Boletus satanas*

Boletaceae

Characteristics: the cap is at first hemispherical, becoming flatter and cushion-like with age and reaching a diameter of 25 cm. The whitish to silver-grey or olive-grey skin does not peel off and it feels velvety when dry. In species of *Boletus* the spores are carried, not on gills, but in tiny tubes below the cap. The openings of the tubes are yellowish in the young fungus, becoming blood-red when fully grown and blue when bruised. The stem is particularly stout and swollen.—**Distribution:** Europe. An uncommon fungus found only in open woodland, particularly under beech trees on calcareous soils, and appearing in July-October.—**Poison:** muscarine. Even small quantities cause poisoning, especially when eaten raw, but fatalities have evidently not been recorded. After a few hours the patient suffers from vomiting and diarrhoea, and will require hospital treatment.

Common Earth Ball *Scleroderma aurantium*

Sclerodermataceae

Characteristics: the fruiting body of this puff-ball is somewhat similar to a potato. The skin is up to 3 mm thick and is whitish-yellow to brownish, fleshy to leathery or corky, with small dark warts. When cut the flesh is pale and fleshy when young, later becoming blue-black and when ripe developing a mass of powdery spores.—**Distribution:** Europe, in deciduous and coniferous woods, particularly under pine trees, but also on heaths and moorland, appearing from July to November, often in large numbers.—**Poison:** not yet identified. The consumption of even small amounts may cause giddiness and fainting. These symptoms, as well as nausea and vomiting, may occur from 15 minutes to 2 hours after the fungus has been eaten; the vision may also be affected.

Yew *Taxus baccata*

Taxaceae

Characteristics: an evergreen plant growing sometimes as a shrub but more commonly seen as a tree with a thick trunk up to 15 m high. Yews grow very slowly and may live for over 1000 years. The bark is at first smooth and red-brown but later becomes grey-brown and splits off in sheets. The leaves are in the form of pointed but soft needles 20-30 mm long and 2 mm across, with a shiny green upper surface and a somewhat paler, matt under surface. The small inconspicuous flowers appear on the undersides of the branches in March-April. The seeds ripen from August onwards. They are enclosed in a fleshy, red cup-shaped aril about the size of a pea.—**Distribution:** Europe, northern and central Asia, North America, extending up to altitudes of at least 1000 m. In Britain commonly seen in churchyards and gardens.—**Poison:** with the exception of the sweet-tasting red flesh of the arils all parts of the plant contain the extremely poisonous taxin as well as other substances. Several cases of poisoning, some fatal, have been reported and many have involved children. Most cases of poisoning are not due to eating the fruit because the hard seeds pass through the alimentary tract undigested, but to chewing the needles and twigs. Yew is not only poisonous to humans but also to animals. For example, horses which have eaten the shoots have been seen to fall down dead within 5 minutes. The poison acts by stimulating the central nervous system, in particular the respiratory centre. About 1-2 hours after eating the leaves the victim starts to vomit and have abdominal pain, giddiness, pupil dilatation, an irregular pulse and finally dies from failure of the respiratory system. In every case the victim must be taken to a medical practitioner.

Common Buttercup   *Ranunculus acris*

Ranunculaceae

Characteristics: a plant 30-70 cm tall, flowering from May to September, with palmate leaves and shiny yellow flowers.—**Distribution:** very common in Europe and northern Asia, particularly in fields and meadows.—**Poison:** saponin, protoanemonin and anemonin, present in all parts of the plant, including the roots. Poisoning of humans is extremely rare but there have been fatal cases due to eating the roots or sucking the plant sap. Irritant substances in the sap can cause dermatitis. The active principles are destroyed by drying, so cattle fed on hay containing buttercup leaves do not usually suffer. The poison also stimulates and then depresses the central nervous system.

Similar poisons are contained in the related Bulbous Buttercup (*R. bulbosus*), Celery-leaved Buttercup (*R. sceleratus*) and Lesser Spearwort (*R. flammula*).

Christmas Rose   *Helleborus niger*

Ranunculaceae

Characteristics: the leafless stems are up to 30 cm tall. Flowers from December to March with flattish, white blooms.—**Distribution:** growing wild in south-eastern Europe, and commonly grown in gardens in Britain and elsewhere.—**Poison:** saponin and the glycosides helleborein and helleborin. The latter affects the skin and mucous membrane while helleborein acts in the same way as digitalin. The symptoms include salivation, abdominal pain, vomiting, colic and diarrhoea. The pulse rate falls and becomes irregular, breathing becomes difficult and finally the victim has convulsions and dies from respiratory failure. Cases of poisoning are rare.

The related Stinking Hellebore (*H. foetidus*) and Green Hellebore (*H. viridis*), both native in Britain, are equally poisonous.

Wood Anemone *Anemone nemorosa*

Ranunculaceae

Characteristics: grows to a height of 10-20 cm and flowers in March to May. Each stem carries a single white or sometimes pale pink flower.—**Distribution:** western and central Europe, north-west Asia and northern America, in deciduous and mixed woodland, under shrubs or in damp meadows. —**Poison:** anemonin and protoanemonin in all parts of the plant. There has been a case of an adult dying after eating 30 fresh plants. The victim suffers from blood in the urine and giddiness, with signs of cerebral irritation, circulatory collapse, respiratory failure and damage to the mucous membrane of the bronchial tree. Medical attention should be sought if a large amount of the plant has been eaten. The poisonous constituents are destroyed by drying.

Baneberry *Actaea spicata*

Ranunculaceae

Characteristics: a perennial plant reaching a height of 30-60 cm. The large leaves are toothed and coarsely divided with long stalks. The small white flowers, carried in a raceme, appear in May to July and are followed by longish-oval berries which are shiny black when ripe; the berries contain a wine-red sap.—**Distribution:** almost the whole of Europe, also northern Asia and northern America, in deciduous and mixed woodland. The plant grows on limestone in some counties of northern England, but is uncommon elsewhere.—**Poison:** protoanemonin in the berries and seeds cause reddening of the skin and blister formation. After ingestion the victim suffers nausea, vomiting, abdominal pain and may experience a sensation of shortness of breath. Very rarely fatal, and evidently not dangerous to animals.

Monkshood *Aconitum napellus*

Ranunculaceae

Also known as Wolfsbane and Aconite.—**Characteristics:** a herbaceous perennial which overwinters as a stout, turnip-like black root. The stem, which is 50-150 cm tall, carries darkish green palmate leaves, each with 5-7 toothed main lobes. The hooded flowers are dark blue to dark violet and they appear from June to September.—**Distribution:** Europe and temperate Asia, in shady places on damp ground, and common in hilly country.—**Poison:** the alkaloid aconitin which is contained in the root, leaves and flowers; the root is particularly poisonous. This has often been described as the most poisonous flowering plant in Europe. Even the mere act of picking the flowers can cause numbness and tingling in the skin, due to penetration of the poison. Fatal poisoning has occurred when Monkshood has been confused with other plants and its leaves have been used in salads. The poison affects the proper conduction of nerve impulses which may result in failure to produce co-ordinated movement. About 10-20 minutes after ingestion the victim has a burning sensation in the mouth, fingers and toes, then over the skin of the whole body, accompanied by sweating. This is followed by a feeling of numbness in the tongue, and an inability to appreciate temperature changes in the skin. He may also suffer from nausea, vomiting, diarrhoea, pains in the head, neck, back and heart region, singing in the ears, disturbances of vision and frequency of micturition. In serious cases of poisoning there may be difficulty in breathing and the pulse may be slow, weak and irregular. Death occurs very suddenly. Medical attention is quite essential.

The yellow *Aconitum vulparia* is also very poisonous and contains the same poisons as *A. napellus*. It has pale green leaves and pale yellow flowers (see the illustration on the right).

Opium Poppy *Papaver somniferum*

Papaveraceae

Also known as the White Poppy.—**Characteristics:** a herbaceous annual growing to a height of 40-150 cm and flowering in June to August. The erect stem carries longish-ovate pale green leaves with toothed edges; their undersides are blue-green. The large flowers are white or greyish-white with a dark spot at the base of each petal. The fruit is an almost spherical capsule up to 5 cm long and 4 cm across, which has several chambers containing the numerous poppy seeds.—**Distribution:** southern Europe, also cultivated in gardens in Britain and elsewhere.—**Poison:** the milky sap contains the alkaloids morphine, codeine, thebaine, papaverine and narcotine. The ripe poppy seeds contain only 0.01% of alkaloid. Opium is obtained from the white milky sap. Poisoning from eating fresh parts has not been recorded. Morphine acts on the central nervous system, relieves pain if present, is a powerful respiratory depressant, decreases glandular secretion and gives a sensation of euphoria. Its peripheral effect is mainly to increase the tonus of smooth muscles. Codeine is a cough depressant. Papaverine has a relaxing action on the smooth muscle of blood vessel walls and increases the tone of alimentary tract musculature. Narcotine enhances the effect of morphine on the central nervous system. Within about an hour of ingestion the victim suffers from headache, a feeling of giddiness, vomiting, increasing numbness, generalized weakness, drowsiness and loss of consciousness. In severe cases breathing becomes irregular, the pulse is weak and the victim falls into a deep sleep.

Greater Celandine *Chelidonium majus*

Papaveraceae

Also known as Swallow-wort.—**Characteristics:** a perennial up to 70 cm tall with an erect hollow stem which is slightly hairy and thickened at the nodes. The leaves are green to greyish-green with pale undersides and are much divided. The yellow flowers are carried in a loose umbel from May to October. Each flower is about 2 cm in diameter. The fruits are pod-like capsules, up to 5 cm long, which contain the ovate black seeds.—**Distribution:** throughout Europe and parts of temperate Asia, particularly on walls, in hedges and waste ground and also in deciduous woods.—**Poison:** all parts are poisonous and contain about 10 different alkaloids. Cases of poisoning are very rare. Externally the poison may produce skin blisters which are easily infected. Internally the poison acts on the whole of the alimentary tract, causing pain, nausea, vomiting, bloody diarrhoea, painful micturition, giddiness, numbness and circulatory disorder. Fatalities are rare.

Rowan *Sorbus aucuparia*

Rosaceae

Also known as Mountain Ash, although not related to the Common Ash.—**Characteristics:** a tree up to 10 m tall. The leaves are pinnate with 9-19 small, ovate, sharply toothed leaflets. The inflorescence is a large corymb with numerous small whitish flowers with a rather unpleasant scent. They appear in May to June. The fruits start to ripen in August, each becoming about the size of a pea. They are at first green, then bright yellow to coral-red.—**Distribution:** in many parts of Europe and parts of northern Asia, particularly in the north, extending almost up to the tree limit in some places. Much planted in parks and streets.—**Poison:** amygdalin, sorbic acid and parasorbic acid in the berries. Cases of poisoning are rare and occur only after consumption of a large quantity of fresh berries. Vomiting and abdominal pain may occur. When well cooked the berries can be made into a safe and palatable jelly.

Common Broom *Sarothamnus scoparius*

Papilionaceae

Characteristics: a shrub, 1-2 m tall, with sparse foliage, which flowers in the spring. The erect, wiry branches carry the small green leaves and bright yellow flowers. The fruit is a brownish to blackish-brown pod containing several seeds.— **Distribution:** western and southern Europe in dry sunny places, and commonly cultivated in gardens.—**Poison:** the alkaloid spartein in the seeds affects the central nervous system, slows the heart and affects the breathing. The symptoms are similar to those in nicotine poisoning but are not so severe. Serious poisoning is not likely to occur in humans or in animals unless large quantities of the seeds are consumed.

Yellow Lupin   *Lupinus luteus*   (not illustrated)
Papilionaceae

Characteristics: an annual plant, 30-60 cm tall, flowering in summer. The long-stemmed leaves have 5-9 finger-like leaflets. The yellow flowers are carried on short stalks in a tall raceme. The fruit is a densely haired pod, up to 6 cm long and 1 cm wide, which contains 4-7 yellowish or reddish, darkly marbled seeds.—**Distribution:** originally southern Europe, but cultivated in many other areas.—**Poison:** various alkaloids which are not destroyed by drying. These substances occur in all parts of the plant but particularly in the seeds and they act on the central nervous system. Soon after eating the seeds the victim suffers from excess salivation, nausea and vomiting, slowing of the heart rate, and depression of the respiration rate. The seeds are sometimes eaten by children because they are rather similar to garden peas. There are also records of animals being poisoned by eating Lupins, but there is considerably variation in their susceptibility.

Lupin   *Lupinus polyphyllus*   (illustrated on right)
Papilionaceae

Characteristics: a perennial plant, up to 1.5 m tall, which flowers from June to August. The leaves consist of 10-15 finger-like, lanceolate leaflets which have silky hairs on the underside. The inflorescence is a tall raceme with 50-80 bright blue flowers.—**Distribution:** originally North America, but now very commonly cultivated in Europe, with numerous cultivars, and may also be found as an escape in waste land.—**Poison:** as for the Yellow Lupin.

The blue *L. angustifolius* is an annual plant about 1 m tall with pale blue flowers.—**Poison:** as for the Yellow Lupin.

Laburnum *Laburnum anagyroides*

Papilionaceae

Characteristics: a shrub or small tree up to 7 m tall, with smooth bark. The pale green, long-stemmed leaves with three leaflets are smooth on the upper surface, hairy underneath. The flowers appear in May-June in long, golden-yellow racemes which hang down from the branches. The seeds ripen in July-August in bean-like pods which are at first green but later become brownish-grey.—**Distribution:** originally southern and south-eastern Europe, but cultivated for several centuries and has been found as an escape as far north as southern Sweden. It occurs mainly in hilly country. On account of its adaptability and attractive flowers Laburnum is much planted in parks and gardens. In some places it has unfortunately been planted close to schools.—**Poison:** cytisin, which occurs mainly in the flowers, seeds and roots. Most cases of poisoning occur among primary schoolchildren who are attracted by the pretty flowers and the seeds. The poison acts first as a stimulant and then as a depressant to the central nervous system. About an hour after ingestion the victim may have a burning sensation in the mouth, nausea, vomiting, a severe thirst, colicky abdominal pain, profuse sweating, headache, circulatory collapse and cramps in muscles. In severe cases death occurs as a result of respiratory failure. It has been reckoned that 3-4 pods with 15-20 seeds would constitute a lethal dose for a child.

Nowadays the cultivar known as *Laburnum vossii* is planted in favour of the ordinary Laburnum as it has longer, more attractive flower racemes. It also has the advantage that most of the seeds do not develop properly and so it is not so dangerous.

Runner Bean *Phaseolus coccineus*

Papilionaceae

Characteristics: a climbing plant with twining stems 2-4 m long. The leaves have long stalks and 3 ovate and entire leaflets. The flowers, which appear between June and September, are bright red. They develop into long, rough-skinned pods which are first green and later brown. Each pod contains 3-5 smooth beans with dark brown markings.—**Distribution:** originally tropical America, but widely cultivated as an annual.—**Poison:** phasin and phaseolunatin. Only the beans are poisonous and then only in the raw state; cooked beans are, of course, quite harmless. About ½-1½ hours after ingestion of the beans the victim suffers from vomiting, diarrhoea, and colicky abdominal pain. Pupil dilatation may be observed. These beans are only dangerous when large numbers have been eaten in the raw state.

French Bean *Phaseolus vulgaris*

Papilionaceae

Characteristics: a smaller plant than the Runner Bean, with similar leaves divided into 3 leaflets which have entire edges. The white to yellowish-white flowers appear in June to September in groups of 6-9. The smooth pods are flattened and at first slightly curved and green, usually becoming yellowish on ripening. Each pod contains 2-8 white seeds 2.5 cm long and 1.5 cm across.—**Distribution:** originally tropical America, but several varieties are now cultivated in Europe and elsewhere.—**Poison:** phasin and phaseolin which, as in the Runner Bean, are destroyed by heat, so the seeds are only dangerous when eaten raw and in large quantities.

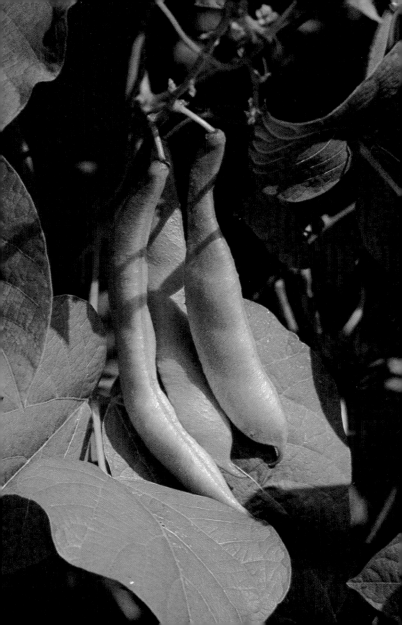

Horse Chestnut *Aesculus hippocastanum*

Hippocastanaceae

Characteristics: a tree up to 20 m tall with a large leafy crown. The dark green leaves are opposite and divided into 5-7 finger-like leaflets, each of which is ovate and toothed. The leaf stalks are up to 20 cm long. During May and June the tree produces a wonderful display of flowers borne in large erect inflorescences. The individual flowers are white or pink to purplish-red. During this period this otherwise rather useless tree produces much nectar which attracts honey bees. In autumn the seeds ripen inside leathery, green, spiny capsules. When fully ripe the capsule splits and the seeds fall to the ground.—**Distribution:** originally Asia, but now introduced and widely planted in many parts of Europe as an ornamental tree in parks, streets and gardens.—**Poison:** the unripe fruits and the green capsules, probably also the shoots, contain saponin. Cases of poisoning from eating these are extremely rare, with only 1-2 recorded cases of death, both in children. Saponin is normally only taken up by the body in small amounts. It acts by breaking down the blood corpuscles (haemolysis). The victim suffers from nausea and vomiting with headache and abdominal pain accompanied by acute thirst, disturbance of vision, dilatation of the pupils, delirium, sleepiness and respiratory failure.

Cypress Spurge   *Euphorbia cyprissias*
Euphorbiaceae

Characteristics: a yellowish-green plant with an erect stem, growing to a height of about 15-30 cm. The leaves are very narrow and somewhat resemble those of a conifer.—**Distribution:** in Europe it occurs in open woodland, often near pine trees, but also on dry heathland and grassland. Known in Britain as a garden plant.—**Poison:** euphorbon, found in the oil in the seeds and also in the milky sap which exudes from any broken part of the plant. The poison has a local action on the skin and mucous membranes where it destroys the tissues, and is particularly dangerous when it enters the eyes. In former days poisoning often occurred as the plant was used as a purgative. Staff in botanic gardens sometimes suffer from the effects of the milky sap, which causes blistering. Taken internally the sap causes burning and reddening in the mouth and oesophagus, and the oil has a purgative action. The victim will vomit, and have abdominal pain, diarrhoea, dilated pupils, giddiness and convulsions followed by delirium, collapse, and rarely death within 2-3 days.

Castor Oil Plant *Ricinus communis*

Euphorbiaceae

Characteristics: an annual shrub growing to a height of about
1-2 m. The stem is thick and brownish-red with large digitate
leaves carried on long stalks. The green flowers appear in
summer, the male flowers in clusters on the upper part of the
plant, the female flowers below. The fruits are spherical capsules about the size of a cherry, which contain the mottled
grey-brown, shiny seeds.—**Distribution:** southern Europe, but
cultivated elsewhere as a pot plant.—**Poison:** the seeds contain the extremely poisonous ricin which was used medically
in the treatment of constipation. When eaten the seeds cause a
sensation of burning, nausea, giddiness, abdominal pain with
watery diarrhoea. In the long term the kidneys may be
damaged and there may be liver dysfunction manifest as jaundice. Death may occur after a few days. As few as 6 seeds are
said to be fatal to a human, but cases are rare and the victims
are usually children.

Mistletoe *Viscum album*

Loranthaceae

Characteristics: an evergreen, perennial plant which lives as a
parasite on the branches of trees, particularly those that are
deciduous, such as Apple. The short, much-branched stems
with yellow-green bark carry longish, leathery leaves which
have no stalk. The white fruits, about the size of a pea, and
with slimy juice, ripen in autumn to spring. Each contain a
single seed.—**Distribution:** Europe and temperate
Asia.—**Poison:** a mixture of acetylcholine and choline, and
the leaves and fruits contain viscotoxin, which acts as a local
stimulant. The victim suffers from vomiting and diarrhoea
with colicky abdominal pain.

Holly *Ilex aquifolium*

Aquifoliaceae

Characteristics: grows as a shrub or tree usually not taller than 6 m, but can attain an age of several hundred years with a trunk 50 cm in diameter and a height of over 10 m. The alternate, evergreen leathery leaves, carried on short stalks, are up to 8 cm long and 4 cm broad. The upperside of the leaves is shiny dark green, the underside pale green. They are elliptical to ovate, usually with a wavy edge armed with sharp spines. The leaves on the upper branches have smooth edges. The small, inconspicuous white flowers, each with 4 petals, appear in May-June in the leaf axils. The spherical to ovate, pea-sized berries, ripen in the autumn. They are at first green, then coral-red. They survive the winter and are still there in the spring, unless eaten by birds.—**Distribution:** Mediterranean area, Europe eastwards to the Caucasus, up to altitudes of 1200 m in the Alps. Holly is frequently grown in gardens, parks and cemeteries.—**Poison:** the alkaloid theobromine occurs mainly in the leaves and berries, in addition to a glycoside. Poisoning mostly occurs among children who have eaten the berries, and fatalities have been recorded. It is said that 20-30 berries constitute a fatal dose. The poison causes intestinal upset with vomiting and diarrhoea. The victim may also become sleepy.

Common Spindle   *Euonymus europaea*

Celastraceae

Also known as Skewer-wood.—**Characteristics:** a shrub
usually about 2-3 m tall, but sometimes more, with a much-
branched stem. The leaves are opposite, dark green above and
pale green below with short stalks. The leaf blade is ovate and
pointed with a finely toothed edge. The small, inconspicuous
yellowish-green flowers appear, usually in May to July, in
small clusters in the leaf axils. These are followed in late summer and autumn by the bright red, four-lobed fruit capsules.
The lobes are enclosed in a fleshy, yellowish-orange aril, each
lobe containing a single seed.—**Distribution:** throughout most
of Europe, in deciduous woodland, along the edges of woods,
in hedges, extending up into the hills in some areas. There are
several cultivated varieties grown in parks and gardens, some
of which have variegated foliage.—**Poison:** the leaves, seeds
and bark contain a bitter substance and also the glycosides
evobioside, evomonoside and evonoside. Poisoning usually
occurs among children who have eaten the very attractive
fruits. The bitter substance causes vomiting and diarrhoea,
while the glycosides act on the heart masculature. The plant
acts as a purgative in animals.

Alder Buckthorn   *Rhamnus frangula*

Rhamnaceae

Characteristics: usually grows as a large erect thornless shrub up to about 4 m in height. The broad leaves are dark green above, pale green below, and have no serrations. they become reddish where the sun strikes them. The whitish-green flowers have 5 petals and 5 sepals. The pea-sized seeds are green when unripe, red when half-ripe and black when fully ripe.—**Distribution:** throughout Europe and parts of temperate Asia, on damp ground, often among scattered birch trees and in open woodland.—**Poison:** the unripe fruits contain saponin, the seeds various glycosides. Ingestion of these parts causes nausea, giddiness, vomiting, intense abdominal pain, water, even bloody, diarrhoea and in serious cases collapse. The effects of the poison are not dangerous, except perhaps to children. Animals can also be poisoned.

Common Buckthorn   *Rhamnus catharticus*

Rhamnaceae

Also known as Purging Buckthorn.—**Characteristics:** usually an erect, much-branched shrub up to c. 2-3 m tall, but it may become a small tree up to 8 m. The leaves are alternate and finely toothed. The small green flowers appear in clusters in the leaf axils in May, June and July. The fruits are at first green and then black when ripe. They have green flesh and a bitter taste. The smaller shoots often end in a stout thorn.—**Distribution:** throughout almost the whole of Europe and parts of temperate Asia, in woodland on calcareous soils.—**Poison:** the seeds and fruits contain the glycosides rhamnoemodine, glucosidorhamnoside, rhamnocathardin, shesterin, while the unripe fruits have saponin. These substances cause a dry mouth and severe thirst as well as vomiting and diarrhoea. Serious poisoning is rare.

Ivy *Hedera helix*
Araliaceae

Characteristics: an evergreen woody plant which climbs with the help of roots along the stem. The dark green, shiny, leathery leaves appear in various forms. Thus, the leaves of the vegetative shrub have 3 or 5 lobes whereas the leaves on the flowering shoots are ovate, without lobes. When the plant has reached a certain age or has grown up above its support it develops branches which no longer climb, but spread out and the leaves are then roundish rather than lobed or angular. The small yellowish-green or whitish flowers are gathered in dense, hemispherical clusters. They are easy to overlook as they are inconspicuous and flower at an unexpected time, namely in October and November. The bitter-tasting, inedible, blue-black pea-sized berries ripen in spring.
—**Distribution:** abundant throughout western and southern Europe, and parts of Asia, in woods, on cliffs and walls and also as an ornamental plant.—**Poison:** the stem, leaves and berries contain saponin, and also hederin. Cases of adults being poisoned are very rare but children who have eaten the berries have suffered serious poisoning, and deaths have been recorded, but only rarely. The berries are hard and taste so bitter that children usually reject them after a trial bite. The ingestion of several berries causes vomiting and diarrhoea. An increase in body temperature may result in convulsions in children and dilatation of the pupils. The sap from the leaves may also cause dermatitis.

Hemlock *Conium maculatum*

Umbelliferae

Characteristics: a biennial or perennial plant, 1-2 m tall, with a spindle-shaped, whitish tap root and an erect, hollow stem, the lower part of which has purplish-red spots. The dark green to grey-green glossy leaves are much divided and carried on round stalks. The small white flowers, appearing in June-September, are borne in umbels with 10-20 rays. The seeds are greenish-brown and ovate, about 3 mm long with wavy ribs. On very hot days this plant emits a repugnant smell reminiscent of mouse urine.—**Distribution:** Europe and parts of Asia, particularly in warmer areas, along the edges of woods, in hedges, extending up into the mountains.—**Poison:** all parts of the plant contain coniine and about 4 other alkaloids as well as ethereal oils. In modern times cases of poisoning due to this plant have been extremely rare and usually the result of confusing it with culinary herbs. There have been cases of children eating the root. In ancient times Hemlock was used in executions. Thus, Plato describes that Socrates died in 399 B.C. after taking a draught of Hemlock. Coniine at first acts by excitation and then depression of the spinal cord and medulla oblongata (part of the brain). About ½-2 hours after ingestion the victim suffers from a burning sensation in the mouth, difficulty in swallowing, excessive salivation, disturbance of the vision and weakness in the limbs. In serious cases the victim may become very weak, with dilated pupils, and there may be vomiting and diarrhoea, with loss of body fluids leading to dehydration. Death is due to respiratory failure.

Cowbane   *Cicuta virosa*

Umbelliferae

Also known as Water Hemlock.—**Characteristics:** a perennial herbaceous plant reaching a height of 60-120 cm. The stem is finely grooved and hollow. The leaves are much divided with sharply toothed, lanceolate leaflets. The white flowers, appearing in July-August, are borne in umbels with 8-12 rays. The seeds are brownish-yellow with distinct dark brown ribs.—**Distribution:** northern and central Europe and parts of Asia, but not extending into the mountains. Grows in marshy ground, ponds, ditches and other wet places, sometimes even half-submerged.—**Poison:** all parts of the plant, but particularly the root, contain the alkaloids cicutoxin and cicutol. Cases of children being poisoned are relatively frequent, probably because they chew the sweet-tasting root. Poisoning has also occurred due to confusion with Celery, Parsnip and the root of Parsley. About 2-3 g of the fresh plant are said to constitute a fatal dose. Even a few minutes after consumption the victim may start to feel a burning sensation in the mouth and throat, together with nausea. This is followed by vomiting, disturbance of vision, fits and there may be loss of consciousness. Finally, death may occur as a result of heart failure or respiratory failure. The plant is said to cause deaths among cattle.

Fool's Parsley   *Aethusa cynapium*

Umbelliferae

Also known as Dog's Parsley and Lesser Hemlock.—**Characteristics:** a herbaceous annual or biennial, 10-100 cm tall. The
hollow stem is finely grooved and much branched. The
smooth, dark green leaves are much dissected to form
feathery leaflets which have an unpleasant smell. The plant
grows between June and October and can be distinguished
from the true cultivated Parsley not only by the unpleasant,
garlicky smell, but also by the leaves which are shiny, particularly on the underside, whereas those of true Parsley are
dull. The flowers are white and borne in flat umbels, while
those of Parsley are greenish-yellow. The spherical to ovate
seeds are 3-5 mm long and pale brown with dark
streaks.—**Distribution:** throughout Europe and parts of temperate Asia, in fields, meadows, railway embankments,
vineyards and as a garden weed.—**Poison:** all parts of the
plant contain aethusin and aethusanol with traces of coniine.
The known cases of poisoning have mostly been due to confusing Fool's Parsley with culinary herbs. The symptoms of
poisoning are similar to those given for Hemlock but less
severe and healthy adults should not suffer unduly. Children
should not be allowed to chew or suck any part of the plant.
Farm animals do not eat fresh Fool's Parsley, presumably
because of its objectionable smell, but evidently it is quite safe
when dry and fed in hay.

Mezereon *Daphne mezereum*

Thymelaeaceae

Characteristics: a shrub 30-150 cm tall, and one of the first to flower in spring. In mild winters the small flowers may appear in February and then on into April. They are pink to pale violet, opening before the leaves appear, and having a powerful sweet scent. The leaves start to appear when the plant is in full flower. They are lanceolate, entire and pale green above, grey-green below. The spherical fruits start to ripen in June-August as pea-sized berries. These are at first green, then bright red and fleshy with a bitter, burning taste. **Distribution:** Europe and parts of temperate Asia, but rare in the north and in low-lying country, preferring hilly country up to altitudes of about 900 m. The plant grows wild mainly in deciduous and mixed woodland, but also under conifers and is much cultivated in gardens.—**Poison:** all parts of the plant contain mezerin and various ethereal oils. Cases of poisoning in children have been recorded, not only from eating the red berries, but also from chewing the shoots. Contact with the skin causes inflammation and the formation of blisters. After ingestion the victim suffers a burning sensation in the mouth, swelling of the lips, excessive salivation, difficulty in swallowing, nausea, vomiting, abdominal pain with diarrhoea and severe thirst. In serious cases delirium will set in and the victim will collapse and die. It is estimated that 10-12 berries will kill a child, unless medical treatment is prompt and efficient.

Guelder Rose *Viburnum opulus*
Caprifoliaceae

Characteristics: a shrub or small tree 2-4 m tall. The broad leaves with 3-5 lobes are similar to those of Maple. The white flowers appear in May-June in loose terminal cymes. The fruits are roundish-ovate berries about the size of a pea, which ripen in late summer. They are then shiny red and hang on the branches until winter.—**Distribution:** throughout Europe along the edges of woodland, on the banks of rivers and lakes and also grown as an ornamental plant in gardens.—**Poison:** the bitter substance viburnin as well as saponin in the fruit, bark and leaves. Fatal poisoning has been recorded. The poison causes irritation of the alimentary tract and the victim suffers from gastroenteritis, vomiting, diarrhoea, giddiness and collapse.

Wayfaring Tree *Viburnum lantana*
Caprifoliaceae

Characteristics: a shrub growing to a height of about 2-4 m. The large ovate leaves have toothed edges and are rough on both sides. The small white, scented flowers appear in dense cymes in May-June. The pea-sized fruits, which are ovate and slightly flattened, ripen from about July to October. At any one time a single cyme may have green, red and black (ripe) berries. The flesh is slimy with a sweetish taste.—**Distribution:** Europe, except in the north, along woodland edges, in open woods and in hilly country, extending in the Alps up to an altitude of 1400 m. Also grown in gardens and parks.—**Poison:** an unidentified glycoside which causes inflammation of the alimentary tract. The victim suffers from vomiting, giddiness and diarrhoea. Not dangerous unless large quantities of berries are eaten.

Fly Honeysuckle   *Lonicera xylosteum*

Caprifoliaceae

Characteristics: a much-branched shrub growing to a height of 1-2 m. The downy leaves are entire, broadly ovate and usually pointed, growing opposite on short stalks. The yellowish-white flowers appear in pairs during summer. The spherical, almost pea-sized, shiny scarlet berries ripen in late summer. They have a bitter, unpleasant taste.—**Distribution:** Europe, frequently among the undergrowth of deciduous and mixed woodland, and along the edges of woods. Also grown in gardens.—**Poison:** xylostein and tanning substances which irritate the alimentary tract. The poison is only dangerous if large amounts of the berries are consumed. The victim suffers severe vomiting and abdominal pains with diarrhoea, red vision, pupil dilatation, photophobia, sweating and cardiac irregularities leading to collapse, respiratory failure and death.

The following honeysuckles have been described in the literature as slightly poisonous, but there is no exact information on their content of poison or on their effect on humans:

Perfoliate Honeysuckle   *(Lonicera caprifolium)*

A climbing shrub up to 4 m tall, growing in hedges and undergrowth along the edges of woods, and also planted in gardens. The dark red pea-sized berries ripen in groups of six.

Black Honeysuckle   *(Lonicera nigra)*

A shrub 60-150 cm tall, growing in shady mountain woods up to altitudes of 1600 m. The blue-black, pea-sized berries ripen in pairs on a long stalk. Not native to Britain.

Common Honeysuckle   *(Lonicera periclymenum)*

A climbing shrub growing to over 5 m occuring in open woodland and hedges. The dark red, pea-sized berries ripen in groups at the end of the shoots.

Snowberry   *Symphoricarpos albus*

Capifoliaceae

Characteristics: a deciduous shrub with thin branches, rarely taller then 2 m. The elliptical, entire leaves have short stalks and the small whitish, bell-like flowers appear in June to August in terminal clusters. The fruits are snow-white, spherical berries, 1.0-1.5 cm in diameter, which ripen in late summer and remain on the branches up to the winter. —**Distribution:** Europe. Much planted in gardens and parks.—**Poison:** the fleshy berries contain saponin and a hitherto unidentified irritant. The external signs of poisoning include irritation of the skin. Ingestion of the berries causes intestinal inflammation with vomiting and diarrhoea. If large amounts of the berries are eaten the victim may become disorientated, and in severe cases even lose consciousness.

Deadly Nightshade *Atropa belladonna*

Solanaceae

Also known as Banewort and Dwale.—**Characteristics:** a shrubby perennial plant with a woody, erect stem, 50-150 cm tall with diverging branches carrying the dark green leaves. The root is thick and fleshy. The rather dull brownish-purple to violet flowers appear in June-August and are followed by berries which are about the size of a cherry. These are at first green, but ripen to glossy black. During the summer months the branches may carry flowers, unripe green and ripe black berries at the same time.—**Distribution:** Europe and parts of central Asia, but rare in the north. The plant grows in open woodland, on waste ground and along the edges of paths. It can be found in certain localities in southern England.-—**Poison:** all parts of the plant, but particularly the berries, contain hyoscyamine and atropine. Poisoning occurs among adults as well as children who eat the sweet-tasting berries. The lethal dose is reckoned as 3-4 berries for children, more than 10 for adults. The poison acts on the central nervous system. The victim suffers flushing of the skin, dilatation of the pupils, becomes agitated and may have hallucinations. The pulse rate and blood pressure increase and there may be disturbance of vision and speech. The victim dies from respiratory failure.

Livestock have also been poisoned by this plant but evidently this is a fairly rare occurrence, and apparently the plant does not affect rabbits in any way. The poison has been known for a long time and even today the constituents are used in medicine. When administered internally and in a carefully calculated dose atropine acts as an antispasmodic.

Henbane *Hyoscyamus niger*

Solanaceae

Characteristics: an annual or biennial growing usually to a height of 40-80 cm, and seldom to 1 m or more. The root is large, thick and turnip-shaped and the stem is erect and hairy. The grey-green leaves are usually unstalked. They are ovate to longish and coarsely lobed. The plant flowers from June to August, sometimes September. The funnel-shaped greenish-yellow flowers are usually purple at the base and the petals have conspicuous violet veins. The flowers are arranged in rows on one side of the stem. The fruit is a globular capsule containing up to 200 small, kidney-shaped seeds. The whole plant has a nauseous smell.—**Distribution:** the Mediterranean region, Europe and western Asia, growing along the edges of paths, on waste land, in stony places and on rubbish heaps.—**Poison:** all parts of the plant, but particularly the root, contain hyoscyamine, atropine and scopolamine. Cases of poisoning are rare and usually due to confusing the root with edible roots or the seeds with beans. The poisons act on the central nervous system. The victim suffers from giddiness and nausea (rarely vomiting), headache, hallucinations, and has a sense of euphoria. The pulse rate and blood pressure increase, the pupils dilate and there is disturbance to vision and speech.

Woody Nightshade *Solanum dulcamara*

Solanaceae

Also known as Bittersweet.—**Characteristics:** a perennial plant with climbing or straggling branches up to about 1.5 m in length; the lower parts of the stem are woody. The stalked dark green leaves vary in shape; the lower leaves are usually lobed, the upper ones heart-shaped. The flowers, which mostly appear in June-August, are purple with brilliant orange stamens and as in other members of the family they have 5 petals and 5 sepals. The oval, pea-sized berries hang in clusters, and become glossy red when ripe; they at first have a bitter taste, but later become sweet. It is not uncommon to find a plant with flowers, unripe (green) and completely ripe berries.—**Distribution:** throughout Europe, also in Asia, in damp woods, thickets and hedgerows. A very common plant in shady places in England and Wales.—**Poison:** the berries and probably also the leaves contain saponin and solanin. The ripe berries are relatively more poisonous than the unripe. The consumption of as few as 6 berries is enough to require medical treatment and it has been reckoned that 10 berries constitutes a lethal dose, but this is a very rare occurrence. The poison acts as an irritant in the mouth and throat, and the victim suffers nausea, vomiting, diarrhoea and giddiness. The pulse rate increases, the pupils dilate and breathing becomes dificult. In most cases the victim recovers in about 24 hours provided he receives proper medical attention.

Black Nightshade   *Solanum nigrum*
Solanaceae

Sometimes known as Garden Nightshade.—**Characteristics:** an annual or biennial herbaceous plant with an erect stem which scarcely exceeds 50 cm in height. The leaves are ovate and usually toothed, with short stalks. The star-like white flowers, with 5 petals and 5 sepals, appear in July-October and are followed by fleshy black berries with a sharp, sour taste. When fully ripe they are glossy black, rarely yellowish-green or red.—**Distribution:** Europe and many other parts of the world, in fields, hedgerows, gardens and waste land.—**Poison:** as for Woody Nightshade.

Potato   *Solanum tuberosum*
Solanaceae

Characteristics: an annual herbaceous plant 40-70 cm tall, with an erect branched stem and a tuberous root. The heart-shaped, entire leaves are hairy on the underside. The flowers, borne in racemes, are usually white with yellow stamens. The fruits ripen as spherical green berries about the size of a cherry. The fruits are inedible, but the tubers developed in the earth are ready for harvesting in summer or autumn, depending upon the variety.—**Distribution:** originally South America, whence the plant was introduced into Europe in the sixteenth century, and is now cultivated throughout the world.—**Poison:** solanin occurs in the flowers, in the green berries, in tubers that have turned green, and also in the shoots and eyes of tubers set to sprout in a dark place before planting. Very small amounts of choline and acetylcholine have been detected in the sap of uncooked potatoes. A few hours after eating the poisonous parts the victim has a burning and irritating sensation in the throat accompanied by a headache. Animals can also be poisoned by eating green potatoes.

Thorn Apple *Datura stramonium*

Solanaceae

Also known as Jimson Weed and Devil's Apple.
—**Characteristics:** an annual herbaceous plant, 30-100 cm tall,
with ovate, pointed, unequally lobed and toothed leaves. The
white to pale violet trumpet-shaped flowers are borne singly.
They are about 10 cm long and appear from June to September. The fruits usually ripen from August to September as
prickly green capsules, rather like those of Horse Chestnut,
and about the size of a hazelnut. When fully ripe the capsule is
brown and it then splits open releasing a large number of
small, kidney-shaped, dark brown seeds. The whole plant has
an unpleasant smell.—**Distribution:** originally native to central America, but introduced to Europe in the sixteenth century, and is now naturalized in parts of Europe and elsewhere.
It grows in waste land, in vineyards, along the edges of
woods, but is not often found in England.—**Poison:** the leaves
and seeds contain the alkaloids hyoscyamine, atropine
and scopolamine. Cases of poisoning have often occurred,
particularly among children who have eaten the seeds, and
these have sometimes been fatal. The poisons act on the central nervous system. The victim becomes restless, may have
convulsions, hallucinations and ataxia, with a rise in pulse
rate and blood pressure, pupil dilatation, disturbance to
breathing, vision and speech. Fatalities have occurred.

Tobacco *Nicotiana tabacum*

Solanaceae

Characteristics: an annual plant up to about 2 m in height with an erect stem, which is not much branched. The lower leaves are broadly elliptical or ovate with short stalks, the upper leaves narrower and usually sessile. Tobacco is usually in flower from June to September. The pink, rarely white flowers are borne in short clusters. The fruit is a capsule containing several pale brown seeds.—**Distribution:** originally central America, whence the plant was brought to Europe by sailors in the sixteenth century. It is now cultivated on an industrial scale in many parts of the world.—**Poison:** all parts of the plant, except the ripe seeds, contain nicotine. The nicotine content of the leaves varies considerably. The leaves also contain other alkaloids. Tobacco poisoning can be very dangerous. Adults are often poisoned when using certain insecticides which may contain a high percentage of raw nicotine. Gardeners, in particular, are likely to inhale the sprayed insecticide and also to absorb it through the skin. After ingestion the victim suffers a burning sensation in the mouth, sweating, headache, giddiness, possible fainting, nausea, vomiting, diarrhoea and abdominal pain. When there is a large intake of poison the victim collapses with respiratory failure and dies within a few minutes. It is reckoned that 50 mg nicotine would be a lethal dose for an adult. This amount would be contained (not in the smoke, but in the tobacco itself) in 1-2 medium-sized cigars or in 2-3 cigarettes.

Foxglove   *Digitalis purpurea*   (figure)
Scrophulariaceae

Characteristics: a biennial or perennial plant growing to a
height of 160 cm. The large, downy basal leaves are ovate with
long stalks. Higher up the stem the leaves become smaller and
narrower with shorter stalks. The numerous flowers, arranged
in a tall raceme, appear in June to August. They are about 4-5
cm long, pink to purple, rarely white, and marked inside with
darker purple spots on a white ground.—**Distribution:**
western and central Europe, in open woodland, forest
clearings and hedgerows. Often grown as a garden
plant.—**Poison:** all parts, but particularly the leaves, contain
digitalin, digitoxin and other related alkaloids. These act on
the heart and on the cardiac and vasomotor centres in the
brain. Chewing the leaves causes irritation of the mouth,
nausea, vomiting. An overdose of a digitalin preparation
causes gastro-intestinal upset, possibly with disturbance of
vision and bradycardia. Refined extracts from this plant have
important medical uses.

Yellow Foxglove   *Digitalis lutea*   (not illustrated)
Scrophulariaceae

Characteristics: reaches a height of about 100 cm and has
narrower leaves than *D. purpurea*. The whitish to pale yellow
flowers are only 2.0-2.5cm long and unspotted.—
Distribution: Europe (rare in the north and east), in
deciduous woods, undergrowth, rocky slopes.—**Poison:** as
for *D. purpurea*.
The related *D. lanata* and the yellow-flowered *D. grandiflora*
grown in gardens contain the same poisons.

Common Bryony *Bryonia dioica* (fig. opposite)
Cucurbitaceae

Also known as White Bryony.—**Characteristics:** a perennial climbing plant reaching a length of 3 m; it climbs by means of tendrils. The matt, pale green hairy leaves with 5-7 lobes resemble those of Ivy. The small, greenish flowers appear in June-July and are succeeded by poisonous red berries about the size of a pea.—**Distribution:** Europe, eastwards to the Caucasus, in hedgerows, undergrowth and along the edges of woods.—**Poison:** bryonin and bryonidin in the berries and root, as well as saponin in the seeds. Consumption of the berries causes nausea, vomiting, colic and watery diarrhoea which may be blood-stained. It is estimated that 15 of the attractive berries constitute a lethal dose for a child, about 40-50 for an adult. Externally the poison causes reddening of the skin, followed by painful inflammation with blister formation. The death rate from consumption of the berries is high.

Bryonia alba (not illustrated)
Cucurbitaceae

Characteristics: a climbing perennial similar to Common Bryony, with greenish-white flowers and spherical black berries about the size of a pea.—**Distribution:** central and southern Europe, in hedgerows, copses and along the edges of woods, but rather rare. Not native in Britain.—**Poison:** the berries and roots contain bryonin and bryonidin. The effects of these substances on humans are the same as in Common Bryony.

Privet *Ligustrum vulgare*

Oleaceae

Characteristics: an evergreen, much branched shrub reaching a height of about 4 m. The lanceolate, opposite, glossy leaves have short stalks and are dark green above, paler below, often becoming violet in autumn. The strongly scented white flowers appear in June and July in terminal panicles. The fruits are glossy black berries about the size of a pea; they have an unpleasant, bitter taste and often remain hanging on the plant until the following spring.—**Distribution:** Europe and western Asia, but rarer in the North. Common in southern England, in open woodland and often planted as a hedge.—**Poison:** the glycoside ligustrin in the leaves and bark, and bitter substances in the bark. Poisoning in children has occurred after consumption of the berries. They cause severe gastro-intestinal upset, with vomiting, diarrhoea and convulsions. The poison also causes irritation of the skin.

Dandelion *Taraxacum officinale*

Compositae

Characteristics: a familiar plant which is in flower from April to autumn. The yellow flowers ripen to a head of seeds each borne on a hair-like structure.—**Distribution:** very abundant throughout Europe, also in northern Asia and America, in fields, meadows, extending up into the hills, and a common garden weed.—**Poison:** the bitter substance taraxin occurs mainly in the stem and root. It causes choleretic, acts as a diuretic, sometimes causes gastro-enteritis and may alter the heart rate. However, this is not a truly dangerous plant, and indeed the dried root has been used as a substitute for coffee.

False Hellebore   *Veratrum album*

Liliaceae

Characteristics: this is a herbaceous plant with an erect, hairy stem which grows to a height of about 1.0-1.5 m and has a tuberous rootstock. The leaves at the base are broadly elliptical, with longitudinal grooves, and they become narrower the higher up they grow; their undersides are downy. The whitish, yellowish or greenish flowers have short stalks and are borne along the tall stem. The related and equally poisonous *Veratrum nigrum* is seldom found in the wild. It has beautiful dark purple flowers and is often grown as an ornamental garden plant.—**Distribution:** central and southern Europe, in damp meadows in mountainous areas, mainly at altitudes between 1000 and 2700 m. Not native to Britain. —**Poison:** all parts of the plant contain the alkaloid protoveratrine as well as other related substances. Poisoning is very uncommon owing to the unpleasant taste of the plant, which was formerly used to prepare a lethal poison and arrow-heads were dipped in the preparation. The alkaloids act on the sensory nerve endings in the skin, eyes, nose and mouth, and also have the effect of reducing the blood pressure. After oral consumption the victim feels pain in the mouth, with burning on the tongue and a prickly sensation in the throat. This is followed by excessive salivation, thirst, vomiting, severe diarrhoea, reduced urine output, muscle cramps and shivering, with a feeling of anxiety. The pulse rate and body temperature decrease, respiration become less effective and collapse may result. death may occur about 3-12 hours after intake of the poison, with consciousness unimpaired.

Meadow Saffron   *Colchicum autumnale*

Liliaceae

Characteristics: a herbaceous plant which during the summer produces a bulb up to about 7 cm across, with a side offshoot, from which the flowers grow in autumn. Then in winter the old bulb dies off and the side shoot forms a new bulb. The plant flowers from August to October. Each plant produces delicate pink to lilac tubular flowers which are rarely taller than 15 cm. When in flower the plants have no leaves. These do not develop until the following spring; they are 20-30 cm long, broad lanceolate with parallel veins. At the same time the fruit capsule develops as a longish-ovate capsule up to 6 cm long, in which the numerous dark brown seeds ripen.—**Distribution:** southern, central and western Europe, in meadows and damp forest clearings, high up into the mountains.—**Poison:** all parts contain colchicine. Several cases of poisoning have been recorded, some of which have been fatal. Most cases involved children who are evidently particularly sensitive to this plant's poison. Poisoning can also occur following the consumption of milk from sheep and goats that have fed on Meadow Saffron. The lethal dose is said to be 20-40 mg (5-10 seeds). The first signs of illness appear 2-6 hours after ingestion, with burning in the mouth, difficulty in swallowing, nausea, frequently vomiting, diarrhoea and colicky abdominal pain. The urine becomes blood-stained. Generalized weakness may lead to respiratory and circulatory insufficiency, and the victim may die in full possession of his faculties.

Common Solomon's Seal

Polygonatum multiflorum (fig. above left)

Liliaceae

Characteristics: a herbaceous plant, 30-70 cm in height with a stout, round stem and alternate ovate to elliptical leaves. In spring the undersides of the stems carry groups of 2-5 whitish, bell-like flowers. The fruits ripen in summer as blue-black berries, about the size of a pea.—**Distribution:** almost throughout Europe and northern Asia, in shady deciduous woods and among shrubs.—**Poison:** a digitalin-type glycoside, mainly present in the berries. Cases of poisoning are rare and would require large numbers of berries. Nevertheless, there is a record of a child dying after eating the berries. The poison acts on heart muscle and its conducting tissue. The victim suffers nausea and abdominal pain with vomiting and diarrhoea. There may also be temporary disturbance to vision.

The Whorled Solomon's Seal (*Polygonatum verticillatum*) has the leaves arranged in whorls and berries which are at first red. It has the same type of poison.

Polygonatum odoratum (fig. above right, and below)

Liliaceae

Characteristics: a herbaceous plant with a curved, angular stem, growing to a height of 15-50 cm. The ovate to elliptical leaves grow in two rows on the upperside of the stem. The sweet-scented whitish, bell-like flowers hang singly, rarely in twos. The berries are blue-black and up to 12 mm in diameter.—**Distribution:** throughout most of Europe, in dry open woods and thickets. Not native to Britain.—**Poison:** as for *Polygonatum multiflorum.*

Lily of the Valley *Convallaria majalis*

Liliaceae

Characteristics: a perennial plant with an underground creeping rootstock which becomes longer each year and produces side shoots. In spring a pair of leaves grow up from the rootstock; they are at first inrolled and later long-stalked. In May and June leafless stems carry the whitish to greenish-white, sweet-scented, bell-like flowers. The spherical, pea-sized red berries hang from August to September on the leafless stems.—**Distribution:** throughout Europe and northern Asia, except the high north and some southern areas, mainly in deciduous and mixed woodland, and in mountain meadows. Frequently grown as an ornamental plant in gardens.—**Poison:** all parts of the plant contain digitalin-type glycosides as well as saponin. Fatal cases of poisoning have been described. Thus, a five-year old child died after drinking water from a flower vase which had contained some Lily-of-the-Valley. Most cases of poisoning are due to eating the red berries, but also to sucking and chewing the leaves and stems. The saponin acts as a local irritant. The victim suffers from nausea, vomiting and colicky abdominal pain. The pulse rate decreases and its rhythm may be irregular. The blood pressure may rise in an attempt to compensate for this, but finally the pulse becomes weak and the victim collapses and dies from heart failure.

May Lily   *Maianthemum bifolium*
Liliaceae

Characteristics: a perennial herbaceous plant, 5-15 cm in height, with two alternate dark green leaves. In May and June the plant carries small white flowers arranged in erect, terminal clusters. These ripen in autumn as small berries which are at first greenish-white, then green with red spots, and finally when fully ripe they are a glossy cherry-red. —**Distribution:** northern temperate zone, frequently in open woodland and along the edges of woods.—**Poison:** digitalin-like glycosides, but there are evidently no recorded cases of poisoning.

Herb Paris   *Paris quadrifolia*
Liliaceae

Characteristics: a perennial herbaceous plant with a leafless stem growing to a height of 25-40 cm, which carries at the top 4, rarely 5 or even 6 horizontally positioned leaves. From May to June these large, unstalked leaves are surmounted by a single, terminal, stalked flower which is greenish-yellow to pale green. The 4-chambered fruit ripens in July and August to form a single blue-black berry, about the size of a cherry, containing numerous seeds. These berries have an unpleasant sweet taste.—**Distribution:** almost the whole of Europe and temperate Asia, more commonly in the south, in shady and damp deciduous and mixed woods, particularly on chalky soil.—**Poison:** paridine and paristyphnine, mainly in the berries and roots. Cases of poisoning have been recorded and one fatality has been described. The victims were mostly children who had confused the berries with the fruits of Bilberry. The poison acts as a local irritant. The victim may also suffer from nausea, vomiting, diarrhoea with colicky abdominal pain, headache, giddiness, constricted pupils and respiratory failure. Medical care is usually only needed if large amounts of the berries have been eaten.

Daffodil   *Narcissus pseudonarcissus*

Amaryllidaceae

Characteristics: a bulbous plant up to 40 cm in height with long, linear leaves which in March-April carries large, trumpet-shaped yellow flowers that are too well known to require further description.—**Distribution:** Europe; rare as a wild plant in open woodland and meadows, but grown in some gardens. There are numerous horticultural varieties (cultivars) of this species.—**Poison:** the bulb contains the alkaloids narcissin and lycorin, and a bitter substance. Cases of poisoning have been known which were due to confusion with onions. Ingestion causes difficulty in swallowing, vomiting, diarrhoea, sweating and possibly convulsions. The sap which oozes from the cut stems has an irritant effect on the skin, and may be troublesome to gardeners.

Poet's Narcissus   *Narcissus poeticus*

Amaryllidaceae

Characteristics: a bulbous plant growing to a height of about 50 cm, with long, linear leaves. The delicately scented white flowers have a yellowish corona with a thin red edge. They appear in April-May.—**Distribution:** Mediterranean region from France to Greece as a wild plant, but widely planted in gardens in Europe and elsewhere, whence some have escaped and become naturalized.—**Poison:** the bulbs contain the alkaloids narcissin and narcipoetin. Here again confusion between the bulbs and ordinary onions has led to cases of poisoning. Ingestion may cause nausea and vomiting, but medical attention will only be necessary if large amounts have been eaten.

Bog Arum *Calla palustris*

Araceae

Characteristics: a marsh plant with a long green, creeping rootstock which produces basal leaves and erect inflorescences. The plant colonizes new ground as the rootstock grows at the tip and dies off at the other end. The leaves are shiny green and heart-shaped with smooth edges and they have long stalks. The outer part of the inflorescence is a broad flat spathe, which is green outside, white inside and marked with conspicuous longitudinal streaks. The spathe partially encloses the spadix, about 3 cm long, which carries the closely packed small green flowers. In summer these ripen into roundish but slightly angular coral-red berries which are sticky to the touch.—**Distribution:** parts of Europe, but not known as a wild plant in Britain. This is a rare plant that grows in low-lying and hilly country in forest swamps, moorland marshes, along the banks of ponds, lakes and slow-flowing rivers.—**Poison:** all parts of the plant contain a substance that is closely related to aroin. Ingestion of the berries may cause poisoning. In former times a concoction made from the root was used as an antidote to snake poisoning. The poison has a local action on the skin and mucous membrane, producing inflammation, sometimes with haemorrhage. After ingestion of the berries the skin may become blistered and the tongue swollen. The victim may suffer abdominal pain, and later cerebral convulsions, with dilatation of the pupils and loss of consciousness.

Cuckoo-pint   *Arum maculatum*

Araceae

Also known as Lords-and-Ladies.—**Characteristics:** a perennial herbaceous plant, 10-25 cm in height, with a tuberous underground rootstock. The long-stalked, arrow-shaped leaves
are dark green, sometimes with blackish spots. The inflorescence, which appears in April-May, consists of a
greenish-white spathe which is enlarged at its base. The spathe
encloses the characteristic spadix with a brownish-red top.
The actual flowers are at the base of the spadix and are shown
in the photograph (bottom, left), where part of the spathe has
been cut away. The seeds develop from the female flowers as
densely packed ovate berries which are at first green and then
shiny red. After the leaves and spathe have withered these
seeds remain for quite a time.—**Distribution:** widespread in
Europe in shady, damp, deciduous woods and under
hedges.—**Poison:** the red berries, the fresh leaves and the
rootstock all contrain aroin. Cases of poisoning have occurred among children who have eaten the sweet-tasting
berries or the rather sour leaves. The poison has a local
irritant action on the skin which may become inflamed even
after just picking the inflorescence. Ingestion causes swelling
of the lips and inflammation of the mucous membrane, with a
sensation of burning. The victim may also suffer hoarseness,
excessive salivation, vomiting and severe abdominal pain, and
the heart rhythm may become irregular. In severe cases the
pupils dilate, fits occur and the victim dies in coma, but such
cases are very rare.

House Scorpion   *Euscorpius italicus*

Scorpionidae

Characteristics: a small scorpion, only 3-4 cm long, with a blackish-brown body. At the front end there is a pair of much enlarged forceps/pincers or pedipalps. The body consists of an anterior part (prosoma) with 7 broad segments and 4 pairs of walking legs, a middle part (mesosoma) with broad segments and a posterior part (metasoma) with 6 narrow segments, forming a very flexible tail. The last body segment carries the large, curved sting which can be erected by bending the rear part of the body, and used to sting prey.—**Distribution:** Mediterranean area. Two other small scorpions, *E. carpathicus* and *E. germanus*, occur in the southern Tyrol and southern Switzerland. *E. carpathicus* has been recorded as far north as Lower Austria.—**Poison:** hyaluronidase. The sting is not dangerous to humans. It causes localized pain at the site of the sting, comparable with the sting of a bee or wasp.

Common Yellow Scorpion   *Buthus occitanus*

Scorpionidae

Characteristics: very similar to the House Scorpion, but about 8 cm long, and yellowish-brown in colour.—**Distribution:** Mediterranean area. Another species, *Scorpio afer*, about the same size, also occurs in the Mediterranean area. As all scorpions are nocturnal animals it is inadvisable to walk around with bare feet after dark.—**Poison:** hyaluronidase and a protein-type toxin. The site of a sting develops local reddening and a sharp prickly pain, followed by numbness. This is followed by a more general reaction with sweating, debility and a decrease in pulse rate. Up to 10 hours after the sting the victim may develop a lung oedema. A doctor should be consulted.

Garden Spider *Araneus diadematus* (fig. above left)
Araneidae

Characteristics: usually yellowish-brown, rarely pale green, always with a conspicuous white cross on the abdomen. This spider spins a large orb web and lies in wait for its prey.—**Distribution:** Europe, where it is common in woodland and meadows, but also spins its web on projecting parts of buildings.—**Poisons:** protein-type toxins. The bite can only cause pain when given in a tender part of the body, which may lead to swelling and local paralysis at the site of the bite.

Water Spider *Argyroneta aquatica* (fig. below)
Agelenidae

Characteristics: a dark grey-brown spider up to about 15 mm long; the male is usually longer than the female. The abdomen is covered with very fine white hairs. The legs are covered with stiff hairs. The spider swims to the surface from time to time in order to renew its air supply. It does this by raising its abdomen above the water where air is trapped by the fine hairs. The abdomen then looks like a large silvery bladder. Each sex builds a separate home shaped like a bell among aquatic plants.—**Distribution:** Europe, in standing and flowing waters with clear water and dense vegetation.—**Poison:** as for the Garden Spider.

Cheiracanthium punctorium (fig. above right)
Clubionidae

Characteristics: a spider about 1 cm long which builds a tubular net in which it catches insects, biting them with the larger venomous chelicerae.—**Distribution:** Germany, France, Switzerland, Italy and Yugoslavia, but not in Britain, which has the smaller *C. erraticum*.—**Poison:** not yet identified. The site of the bite is painful and suffers swelling and becomes bluish-red. The victim may suffer from nausea, vomiting, headache and a slight increase in body temperature. These symptoms subside after about 3 days, but the bite may remain red and swollen for longer.

Malmignatte *Latrodectus tredecimguttatus*
Theridiidae

Characteristics: a small spider, the body only about 7-10 mm long, with long legs. The body is deep black, the abdomen being marked with 13 characteristic red spots.—**Distribution:** Mediterranean area. Most bites from this spider occur out in the open, but it may also enter bathrooms, and hide behind mirrors or inside locks.—**Poison:** protein-type venoms. Within 10 minutes the bite becomes very painful. This is followed by swelling of the lymph nodes, a rise in blood pressure, difficulty in breathing, sometimes with facial sweating, speech disturbance and paralysis of the jaw muscles. The abdominal muscles may also become extremely hard. Loss of appetite and insomnia may occur. These bites should be treated by a doctor.

Tarantula *Lycosa tarentula*
Lycosidae

Characteristics: a brownish-grey spider, the body length of 3-5 cm, and with dark transverse bars on the abdomen. The spider spends the day in holes in the ground down to a depth of 20-25 cm and during the night comes out to catch prey, chasing mainly on the ground. Only during spring is it also active by day.—**Distribution:** Mediterranean area, mainly in Italy, Sardinia and Spain.—**Poison:** not yet identified. It is cytotoxic. The old idea that the victim is stimulated to dance the tarantella before he or she can recover from this spider's bite is, of course, completely erroneous. The bite is not dangerous nor very painful and in fact no worse than a bee sting.

Castor Bean Tick   *Ixodes ricinus*

Ixodidae

Characteristics: ticks and mites are not insects, but are more closely related to spiders. This species has eight legs and feeds on the blood of mammals, occasionally even on the blood of humans. The adult male is about 2 mm long, the female 4 mm, but they may reach a length of 10 mm when gorged with blood. Ticks have a biting proboscis with hooks.—**Distribution:** many parts of Europe, and elsewhere, in meadows, woods and other damp places.—**Poison:** strictly speaking this tick is not a poisonous animal, but it transmits to man the virus of a form of meningo-encephalitis. There may be fatalities or the condition can become chronic (headache, depression). The disease also leaves behind a lasting immunity. About 3-4 weeks after the bite, specific antibodies can be demonstrated in the victim's blood. The clinical picture has two phases: 1. after an incubation period of 3-14 days the victim suffers headaches, fever and pains in the abdomen and limbs. These symptoms usually remit within 4-6 days and the condition is resolved. 2. In other cases (8-10% of infected persons) a second phase ensues after 4-6 days without pain. This is characterized by high fever, severe headache, vomiting and a weakness of certain muscles, particularly those of the shoulder girdle. Although the tendency to remission is good there are always some fatalities. The ticks are particularly active in May-June and September. These are the times to avoid their haunts and not to sit down for a rest along the edges of woods or in forest clearings. After a walk the skin should be carefully searched or ticks, and this applies particularly to young children. Attached ticks should be removed, but without using force. It is best to cover the affected area with a thick layer of oil, fat or skin cream. The ticks should then fall off; in some cases warm water will be sufficient. A drop of glue that is not water soluble is also effective. After this the bitten person should keep a look-out for the symptoms described above, and if they appear should consult a doctor.

It would be wrong, however, to think that these ticks are ubiquitous. They occur mainly among damp vegetation and cannot thrive in dry conditions. They move up to the top of a plant and remain there until they feel the touch of a passing animal. They then attach themselves firmly and move to a

suitable place for sucking blood. They remain on a host for 4-6 days and during this time become much enlarged with blood. They then release their grip, fall to the ground and start to digest the blood. Young ticks have to have two meals of this type and the female has to find a third host and suck more blood before she lays her eggs—over 2000 of them.

Common Water-boatman   *Notonecta glauca*

Notonectidae

Characteristics: a water bug about 15 mm in length. This is one of the back-swimmers, so called from their habit of swimming upside-down and "hanging" in this position from the water surface. It is lighter than water and when it comes to the surface with its back upwards it is able to fly off immediately. The arched back is paler than the belly. The bug swims with its long hind limbs which are fringed with swimming hairs. Prey is held with the front limbs and its body contents are sucked out with the proboscis.—**Distribution:** Europe, in standing waters.—**Poison:** unidentified substances. The bite is painful but not dangerous.

Water Scorpion   *Nepa cinerea*   (fig. below, left)

Nepidae

Characteristics: a grey-brown aquatic bug, about 2 cm long with a longish-ovate flat body and usually with a layer of mud on the back. The middle and hind pairs of legs are used for walking and swimming, but the two front legs are modified for catching and holding prey, from which the bug sucks the contents through its proboscis. The needle-like structure at the rear end of the body is the respiratory tube.—**Distribution:** Europe, almost exclusively in slow-flowing or standing waters, mainly in shallow places near the banks. They mostly lie in wait for prey on the bottom or perched on water plants. —**Poison:** as for the Common Water-boatman.

Rhinocoris iracundus

Reduviidae

Characteristics: a red and black predatory bug about 12 mm long with a narrow head which is much restricted behind the eyes. The bug uses its short, powerful proboscis to pierce, paralyse and kill prey.—**Distribution:** Europe, but not in Britain, usually sitting on plants lying in wait for insects to alight.—**Poison:** as for the Common Water-boatman.

122

Cleg *Haematopoda pluvialis*

Tabanidae

Characteristics: a small, inconspicuous, grey to blackish-grey horse fly, about 10 mm long, with marbled wings. The two large compound eyes (fig. above, right) are particularly characteristic, those of the male meeting in the middle, those of the female separated from one another by a narrow gap. The attractive eye clolours of horse-flies are often used in identification. The mouth parts are modified for piercing. Those of the female can penetrate human skin and suck blood. The males feed only on plant saps.—**Distribution:** widely distributed throughout Europe, particularly in damp areas near to rivers and lakes, where they often occur in enormous numbers—**Poison:** the saliva contains a substance which inhibits coagulation of the blood. This substance causes itching and swelling at the site of the bite.

Mosquito *Theobaldia annulata*

Culicidae

Characteristics: a large, inconspicuously coloured mosquito, 8-10 mm in length, with a slender body and long, thin legs. The males feed on plant sap but the females suck blood, which is necessary for the development of the eggs. When the mouth parts of a female penetrate human or other skin an anticoagulant saliva flows into the wound. If the insect is disturbed during this process considerably more of this substance is produced and the mosquito then flies off. —**Distribution:** throughout Europe, mainly in the vicinity of water, where they may occur in large numbers. In summer they are active in the morning and evening but also during the day when the sky is overcast.—**Poison:** substances that inhibit blood clotting. At the time the sting is scarcely felt but the site soon starts to itch and produce a weal. This unpleasant itching sensation can last for at least a day.

Honey-bee *Apis mellifica*

Apoidea   (fig. below is Bumble-bee, text on p. 128)

Characteristics: a dark brown insect with a somewhat hairy body, transparent wings and at the rear end in the females only a venomous sting. The sting is barbed and normally cannot be withdrawn. It remains in the wound together with the venom gland and the bee dies. The somewhat larger drones or males have no sting.—**Distribution:** all parts of the world except the very cold regions.—**Poison:** mellitin, histamine, phospholipase and hyaluronidase. Mellitin, which forms the major part of the bee's poison, has a haemolytic action, but it can also affect the heart. Histamine is a pain-producing substance which acts on the circulatory system by lowering the blood pressure. Phospholipase and hyaluronidase are proteins and may function as antigens. In humans each sting yields about 0.1 mg of poison (calculated as dry weight). About five bee stings cause mild poisoning, 40 stings produce serious symptoms, but 500 stings probably constitute a lethal dose. In hypersensitive persons, however, even a single sting can lead to death due to anaphylactic shock. In the area of the sting there may be a burning sensation, itching, inflammation, swelling and acute pain. Great care should be taken not to swallow a bee when partaking of a sweet dish or drink. In such a case the rapid swelling of the mucous membrane of the throat may lead to suffocation. Bee stings may, however, produce more general symptoms, including nausea, possibly vomiting, a rise in body temperature to 39.5°C, headache, difficulty in breathing and a decrease in blood pressure. The sting should be carefully removed with forceps, taking care not to press on the attached venom gland.

Bumble-bees   *Bombus* species   (fig. p. 127, below)
Apoidae

Characteristics: an insect 12-20 mm long with a very hairy body and bright colours.—**Distribution:** throughout most parts of the world. There are several species, and about ten in Europe.—**Poison:** serotonin, phospholipase A and hyaluronidase. Serotonin acts as a pain-producer and may affect blood pressure. The other substances act as described for the Honey-bee.

Wasps   *Polistes* and *Vespa* species   (fig. opposite, above)
Vespoidea

Characteristics: most wasps reach a length of 15-25 mm and have a body with a striking black and yellow pattern. Only the females (queen and workers) have a sting, which is not barbed and can therefore be withdrawn).—**Distribution:** numerous species in all parts of the world.—**Poison:** histamine, serotonin, wasp-kinin, phospholipase and hyaluronidase. The symptoms produced are similar to those described for the Honey-bee.

Hornet   *Vespa crabro*   (fig. opposite, below)
Vespoidea

Characteristics: similar to an ordinary wasp, but much larger, up to 30 mm long.—**Distribution:** widespread, but not in large numbers.—**Poison:** histamine, serotonin, acetylcholine, hornet-kinin, phospholipase. In general, the effects of a sting are similar to those described for the Honey-bee. The stings of several Hornets can kill a child owing to heart and respiratory failure.

Fire Salamander *Salamandra salamandra*

Salamandridae

Characteristics: an elongated black amphibian, up to 20 cm in length (up to 28 cm in southern Europe), with irregular bright yellow to orange markings. This is a slow-moving rather plump animal with a relatively broad head. The smooth, slimy skin has several gland pores.—**Distribution:** widespread in central and southern Europe, including Corsica and the Iberian peninsula. Rare in low-lying country, but usually in shady forest clearings in hilly and mountainous country, extending up to altitudes of about 1000 m in the Alps.—**Poison:** the skin secretions contain several alkaloids, which act on the central nervous system and as an irritant to the mucous membrane. In general, humans only suffer if they handle these animals and then rub their eyes, without having washed their hands.

Alpine Salamander *Salamandra atra*

Salamandridae

Characteristics: more slender than the Fire Salamander and reaching a length of about 16 cm. The body is shiny black. There is a large gland on each side of the head in the ear region, and the sides of the body have a distinct ribbed appearance with protruding warts.—**Distribution:** only in the Alps and mountains of the western Balkans up to an altitude of over 3000 m. Completely independent of water, this salamander lives in damp woods and above the tree line even in alpine meadows.—**Poison:** see Fire Salamander.

Fire-bellied Toad *Bombina bombina*

Discoglossidae

Characteristics: a small, relatively slender toad, up to about 5 cm in length. The upperside is greyish with darker markings and numerous small, roundish warts. The belly is bluish-black with large red markings and white dots. The males have internal vocal sacs. When disturbed these toads turn on to their backs, stretch up their legs and thus expose the warning coloration to an intruder.—**Distribution:** eastern Europe, in ponds and ditches.—**Poison:** a secretion from the skin glands, which acts as an irritant to the mucous membrane of the eyes, nose and mouth. There should be no further symptoms, provided the hands are washed after handling these toads. It has been observed that when alarmed they produce so much secretion that they appear to be covered in a soapy foam and smell of onion.

Yellow-bellied Toad *Bombina variegata*

Discoglossidae

Characteristics: a rather thick-set toad, about 5 cm in length. The back is olive-grey to grey-brown with numerous small warts. The belly is blue-grey with large bright yellow markings. In contrast to the Fire-bellied toad the males of this species do not have vocal sacs.—**Distribution:** central and southern Europe, but not in the Iberian Peninsula and Sicily, usually in hilly or mountainous country up to altitudes of over 1500 m, and only found in water.—**Poison:** see Fire-bellied Toad.

Green Toad *Bufo viridis*

Bufonidae

Characteristics: up to 9 cm in length. The upperside is pale grey to olive-coloured with greenish markings and reddish warts. Using the long hind legs this toad can jump well and rapidly.—**Distribution:** central and southern Europe, extending north to Denmark, but absent from Britain, Belgium, Netherlands and the Iberian Peninsula. A very adaptable species, often found in fairly dry habitats.—**Poison:** bufotenidin and bufoviridin. These irritate the skin and raise the blood pressure. There is no danger, but hands should be washed.

Common Toad *Bufo bufo*   (fig. below, left)

Bufonidae

Characteristics: a squat toad with a broad rounded head. Males are up to 8 cm, females up to 13 cm in length. The upperside may be grey-brown, red-brown or blackish-brown with large, closely packed warts. Behind each ear there is a large halfmoon-shaped gland.—**Distribution:** central and northern Europe, in open woodland, undergrowth, gardens, even entering buildings, particularly cellars.—**Poison:** bufotoxin, bufotalin, bufotenin in glandular secretions. These irritate the mucous membrane of the mouth, nose and eyes, so hands should always be washed after handling these toads.

Common Tree Frog *Hyla arborea*

Hylidae   (fig. below, right)

Characteristics: up to 5 cm in length, with a bright green back, but the colour can change to grey or brown. The fingers and toes have small adhesive pads which enable this frog to climb high up in trees and bushes. When croaking, the males have a large vocal sac which becomes spherical when inflated.—**Distribution:** central and southern Europe (not Britain), in damp meadows, woodland edges and gardens.—**Poison:** a haemolytic peptide of unknown composition which is not dangerous, but the hands should be washed after handling these frogs.

134

Adder or Common Viper *Vipera berus*

Viperidae   (fig. above, female; below, male)

Characteristics: the male is scarcely ever longer than 60 cm, but the female may reach a length of over 80 cm. The head is scarcely set off from the thick-set body. The males are silver-grey, ash-grey or brownish-grey with a black zigzag band along the back. The females are yellowish-brown or red-brown with a dark brown dorsal band. Completely black individuals are not uncommon (fig. p. 138 above), and there is a rarer coppery form (fig. p. 138 below), in which the dorsal pattern is scarcely discernible. The venomous fangs are similar to those of the Nose-horned Viper.—**Distribution:** most of Europe and extending north at least to the Arctic Circle. Also across northern Asia to the Pacific Coast. On heaths, moors, marshy meadow, and in open woodland and hedgerows.—**Poison:** see Asp Viper (p. 140). The bite is painful; the skin punctures made by the two fangs are about 1 cm apart. The site of the bite becomes swollen and bluish-red. The pain gives way to an itching sensation which may last for 8-10 days. In some cases there may be inflammation of the lymph vessels and lymph nodes, usually in association with general symptoms, which include giddiness, headache, nausea, vomiting, diarrhoea, haemorrhage of mucous membranes, rapid thready pulse, decrease in blood pressure and respiratory failure. Bites on the face, neck and back are dangerous, particularly in children. Fewer than one per cent of cases are fatal, and 50 per cent show no general symptoms.

Orsini's Viper *Vipera ursinii*   (not illustrated)

Viperidae

Characteristics: a relatively slender snake, reaching a length of 50 cm at the most, with a narrow head and scarcely any neck. The back is pale grey or pale brown with a dark longitudinal zigzag stripe.—**Distribution:** scattered populations in south-eastern France, Italy, Austria and the Balkans.—**Poison:** see Asp Viper (p. 140).

Nose-horned Viper *Vipera ammodytes*

Viperidae

(fig. p. 141, above)

Characteristics: the largest and most dangerous snake in Europe. The female reaches a length of up to 90 cm, the male is somewhat smaller. The body is thick-set and clearly set off from the broad, triangular head. At the tip of the snout there is a quite characteristic scaled nose-horn, which is not seen in any other European viper. In the male the back is usually grey or grey-brown with a very dark, sometimes even black zigzag band. In the female, which is more brownish, the zigzag band is not so clearly differentiated from the ground coloration. The belly is yellowish with dark grey spots. This is a livebearing viper which feeds on mice, moles, small birds and lizards. The prey is killed by bite before being swallowed. All the European vipers have a fang on each side of the upper jaw with which they inject venom from the venom glands.
—**Distribution:** south-eastern Europe, with the range extending north-westwards as far as Venice, southern Tyrol and Carinthia.—**Poison:** neurotoxins, haemorrhagic agents, proteolytic enzymes, phospholipase, hyaluronidase. Bites may be fatal, particularly when they strike a blood vessel. The venom paralyses the respiratory system, and breaks down blood and tissues. The bite itself is not very painful but it usually bleeds: the punctures left by the two fangs are about 6 mm apart. However, the bite soon becomes painful, swollen and red, while the victim becomes pale. In severe cases the lymph nodes swell and the general condition of the victim deteriorates with irregularities of the pulse, rapid shallow breathing. After some hours the area round the bite shows reddish-blue to dark violet spots as blood leaks out into the tissues. If not treated the tissues become necrotic and a limb may be lost. When a bite pierces a blood vessel death may result when the venom reaches all parts of the body very rapidly. The victim should be kept quiet and taken immediately for medical care.

◁Adder or Common Viper, black colouring (above) and coppery (below). See p. 136.

139

Asp Viper   *Vipera aspis*   (fig. opposite, below)
Viperidae

Characteristics: a thick-set viper with a broad, triangular head
not markedly set off from the body. Females may reach a
length of 75 cm, males are somewhat shorter. The tip of the
snout is distinctly upturned. The back may be grey, pale
brown or red-brown with a dark, almost black pattern, con-
sisting of numerous transverse bands or rectangular markings
which more or less form a zigzag band. The fangs are similar
to those in the Nose-horned Viper.—**Distribution:** France,
Switzerland and Italy, and extending to an altitude of over
2400 m in the Pyrenees.—**Poison:** agents causing
haemorrhages, also phospholipase, and anti-coagulants. In
general, the bite is more dangerous then that of the Adder.
The venom acts on the blood and body tissues and damages
the blood vessels. Haemorrhages occur in a wide area round
the bite and within a few hours may lead to necrosis. The victim may vomit and feel giddy, his temperature may rise and
his pulse rate fall. In severe cases there may be damage to the
kidneys and liver.

INDEX

English Names

Latin Names

Fungi

Amanita muscaria 20
  pantherina 22
  phalloides 16
  virosa 19
Boletus satanas 30
Clavaria formosa 14
Clavaria pallida 14
Gyromitra esculenta 14
Inocybe patouillardii 22
Lactarius torminosus 28
Paxillus involutus 28
Ramaria formosa 14
Rhodophyllus sinuatus 24
Russula emetica 26
  queletii 26
Scleroderma aurantius 30
Tricholoma pardinum 24

Flowering Plants

Aconitum napellus 38
  vulparia 38
Actaea spicata 36
Aesculus hippocastanum 52
Aethusa cynapium 70
Anemone nemorosa 36
Arum maculatum 112
Atropa belladonna 80
Bryonia alba 94
  dioica 94
Calla palustris 110
Chelidonium majus 42
Cicuta virosa 68
Colchicum autumnale 100
Conium maculatum 66
Convallaria majalis 104
Daphne mezereum 72
Datura stramonium 88
Digitalis grandiflora 92
  lanata 92
  lutea 92
  purpurea 92

Euonymus europaea 60
Euphorbia cyprissias 54
Hedera helix 64
Helleborus foetidus 34
Helleborus niger 34
Helleborus viridis 34
Hyoscyamus niger 82
Ilex aquifolium 58
Laburnum anagyroides 48
Ligustrum vulgare 96
Lonicera caprifolium 78
  nigra 78
  periclymenum 78
  xylosteum 76
Lupinus angustifolius 46
  luteus 46
  polyphyllus 46
Maianthemum bifolium 106
Narcissus poeticus 108
  pseudonarcissus 108
Nicotiana tabacum 90
Papaver somniferum 40
Paris quadrifolia 106
Phaseolus coccineus 50
  vulgaris 50
Polygonatum multiflorum 102
  odoratum 102
  verticillatum 102
Ranunculus acris 34
  bulbosus 34
  flammula 34
  sceleratus
Rhamnus catharticus 62
  frangula 62
Ricinus communis 56
Sarothamnus scoparius 44
Solanum dulcamara 84
  nigrum 86
  tuberosum 86
Sorbus aucuparia 44
Symphoricarpos albus 78
Taraxacum officinale 96

Taxus baccata 32
Veratrum album 98
  nigrum 98
Viburnum lantana 74
  opulus 74
Viscum album 56

Animals

Apis mellifica 126
Araneus diadematus 116
Argyroneta aquatica 116
Bombina bombina 132
Bombina variegata 132
Bombus—species 128
Bufo bufo 134
Bufo viridis 134
Buthus occitanus 114
Cheiracanthium punctorium 116
Euscorpius carpathicus 114
  germanus 114
  italicus 114
Haematopoda pluvialis 124
Hyla arborea 134
Ixodes ricinus 120
Latrodectus tredecimguttatus 118
Lycosa tarentula 118
Nepa cinerea 121
Notonecta glauca 112
Polistes—species 128
Rhinocoris iracundus 122
Salamandra atra 130
  salamandra 130
Scorpio afer 114
Theobaldia annulata 124
Vespa—species 128
Vipera ammodytes 139
  aspis 140
  berus 136
  ursinii 136